Harald H. Kessler (Ed.)
Molecular Diagnostics of Infectious Diseases

Also of Interest

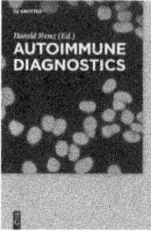

Autoimmune Diagnostics
Harald Renz (Ed.), 2012
ISBN: 978-3-11-022864-9, e-ISBN: 978-3-11-022865-6

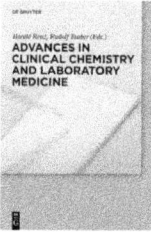

Advances in Clinical Chemistry and Laboratory Medicine
Harald Renz, Rudolf Tauber (Eds.), 2012
ISBN: 978-3-11-022463-4, e-ISBN: 978-3-11-022464-1

Clinical Chemistry and Laboratory Medicine (CCLM)
Published in Association with the European Federation of Clinical
Chemistry and Laboratory Medicine (EFLM)
Editor-in-Chief: Mario Plebani
ISSN: 1437-4331

www.degruyter.com

Molecular Diagnostics of Infectious Diseases

Edited by
Harald H. Kessler

3rd fully revised edition

DE GRUYTER

Editor
Prof. Dr. med. Harald H. Kessler
Medizinische Universität Graz
Zentrum für Angewandte Biomedizin
Universitätsplatz 4
8010 Graz, Österreich
E-Mail: harald.kessler@medunigraz.at

ISBN 978-3-11-032788-5
e-ISBN 978-3-11-032812-7
Set-ISBN 978-3-11-032813-4

Library of Congress Cataloging-in-Publication data
A CIP catalog record for this book has been applied for at the Library of Congress.

Bibliographic information published by the Deutsche Nationalbibliothek
The Deutsche Nationalbibliothek lists this publication in the Deutsche Nationalbibliografie; detailed
bibliographic data are available in the Internet at http://dnb.dnb.de.

© 2014 Walter de Gruyter GmbH, Berlin/Boston

Typesetting: Compuscript Ltd.
Printing and binding: CPI buch bücher.de GmbH, Birkach
Cover image: Luchschen, Getty Images/iStockphoto

♾ Gedruckt auf säurefreiem Papier
Printed in Germany
www.degruyter.com

Preface to the Third Edition

A period of only two years elapsed between the appearance of the first and second editions of this book. This third edition has been produced after just another two years since almost all copies of the second edition of this book were sold within slightly more than a year.

Interesting and exciting developments have occurred since the publication of the second edition that made it appropriate to revise portions of the book and to present additional material of interest to the readers. All of these new findings and achievements are reflected in this third edition while the themes that initially inspired the creation of this book were continued. The book maintains its tradition of clearly arranged chapters with an emphasis on clinical practicality along with a detailed review of the technical background of methods described. In the general part (chapters 1–6), preanalytical, analytical, and postanalytical issues are discussed. Furthermore, special attention is drawn on quality assurance/quality control and standardization issues. In the special part (chapters 7–13), nucleic acid testing for detection of pathogens producing infectious diseases is discussed in detail. Each chapter focuses on infectious diseases targeting a specific body tract or system. Throughout the text, practice points are found that highlight clinical scenarios requiring targeted molecular diagnostics.

In this third edition, the whole text has been carefully reviewed; all chapters have been updated and several new sections added including those about standardization of molecular assays and the polyomavirus JC as another pathogen relevant in immunosuppression. Furthermore, new tables and figures have been added. All authors and contributors have been commissioned to ensure that the material is fresh, up to date, and relevant.

The overall philosophy and approach of the book is unchanged. As with the previous edition, the standard style throughout the text was maintained. I hope that the uniformity achieved will make the text appreciably easier to read and assimilate. Altogether, this book can be used as a starting point when one needs to evaluate which molecular method, test system, or instrument may be considered or chosen for diagnostics of infectious diseases.

I wish to acknowledge outstanding assistance from all authors and contributors in the preparation of this book, without whom the third edition of this book would not have been possible. I would especially like to extend my sincere gratitude to the very helpful staff at De Gruyter for smoothing the path towards publication. I trust that the readers will enjoy this new edition and benefit from the additional material. While this book tries to cover the whole field and to provide a maximum timely content, the reader should always consider that this field is still changing rapidly. Suggestions for further improvements to be considered in a future edition are highly appreciated. Please contact the editor at harald.kessler@medunigraz.at.

Harald H. Kessler

Contents

Preface —— v

Authors Index —— xiii

Contributors Index —— xv

1   Choice of adequate sample material —— 1
         Holger F. Rabenau, Reinhard B. Raggam, Margit Hübner and Eva Leitner
1.1   Viruses —— 1
1.2   Bacteria —— 19
1.3   Fungi —— 25
1.4   Protozoa —— 27

2   Stability of the specimen during preanalytics —— 29
         Georg Endler, Georg Slavka and Markus Exner
2.1   Sample integrity during collection —— 29
2.1.1   Blood —— 29
2.1.2   Urine —— 30
2.1.3   Stool —— 30
2.2   Degradation of DNA —— 30
2.3   Degradation of RNA —— 31
2.4   Inhibitors of PCR —— 32
2.5   How can contamination during specimen
             collection and in the laboratory be avoided? —— 33
2.6   How can the sample identity be ensured? —— 34
2.7   Transport of diagnostic material —— 34
2.7.1   *Category A Infectious Substances* —— 34
2.7.2   *Category B Infectious Substances* —— 35
2.7.3   *Exempt patient specimens* —— 36
2.8   Stability of nucleic acids of selected pathogens
             during preanalytics —— 36
2.8.1   Human immunodeficiency virus type 1 (HIV-1) RNA —— 36
2.8.2   Hepatitis B virus (HBV) DNA —— 37
2.8.3   Hepatitis C virus (HCV) RNA —— 38
2.8.4   *Chlamydia trachomatis* and *Neisseria gonorrhoeae* DNAs —— 38
2.8.5   Viral pathogens producing respiratory tract infections —— 39
2.8.6   Pathogens in stool specimens —— 39
2.9   Take home messages —— 40
2.10   Further reading —— 40

3   **Quality assurance and quality control —— 41**
Reinhard B. Raggam, John Saldanha and Harald H. Kessler
3.1   Accreditation issues —— 41
3.2   Standardization of diagnostic tests or test systems —— 42
3.3   Validation and verification work —— 44
3.4   Components of validation work —— 44
3.4.1   Internal and external quality controls —— 44
3.4.2   Proficiency testing —— 47
3.4.3   Validation of employee competency —— 48
3.4.4   Instrument maintenance and calibration —— 48
3.4.5   Correlation with clinical findings —— 49
3.5   Components of verification work —— 49
3.5.1   Components of verification work for IVD/CE labeled and/or FDA-approved or -cleared tests or test systems —— 50
3.5.2   Components of verification work for laboratory-developed tests or test systems —— 53
3.6   Take home messages —— 54
3.7   Further reading —— 55

4   **Extraction of nucleic acids —— 57**
Harald H. Kessler
4.1   Manual nucleic acid extraction protocols —— 57
4.2   Automated nucleic acid extraction platforms —— 58
4.2.1   Technology principle —— 58
4.2.2   Desirable features of automated platforms —— 59
4.3   Preparation of qPCR mixes and addition of eluates (qPCR assay setup) —— 60
4.4   Currently frequently used commercially available platforms —— 60
4.5   Take home messages —— 62
4.6   Further reading —— 62

5   **Amplification and detection methods —— 63**
Stephen A. Bustin and Harald H. Kessler
5.1   Nucleic acid-based tests —— 64
5.2   Target amplification methods —— 65
5.2.1   Real-time polymerase chain reaction (qPCR) —— 66
5.2.2   Isothermal amplification techniques —— 75
5.2.3   Next generation sequencing (NGS) —— 78
5.3   Signal amplification methods —— 79
5.3.1   Branched DNA (bDNA) —— 80
5.3.2   Hybrid capture assay —— 80

5.4 What are the key challenges for the future? —— 81
5.5 Take-home messages —— 82
5.6 Further reading —— 83

6 **Interpreting and reporting molecular diagnostic tests —— 85**
Ranjini Valiathan and Deshratn Asthana
6.1 Detection of viral infections —— 85
6.2 Detection of bacterial infections —— 86
6.3 Quantitative endpoint PCR —— 87
6.4 Real-time PCR (qPCR) —— 88
6.5 Reporting results —— 89
6.5.1 Genetic names —— 91
6.5.2 Recommendations for reporting results
       of molecular tests —— 91
6.5.3 Recommendations for the contents of the molecular
       test report —— 92
6.6 Interpretation —— 93
6.7 Important issues when clinically interpreting molecular
       diagnostic results —— 95
6.8 Take home messages —— 96
6.9 Further reading —— 96

7 **Human immunodeficiency virus —— 97**
Jacques Izopet
7.1 Major symptoms —— 99
7.1.1 Untreated individuals —— 99
7.1.2 Treated individuals —— 100
7.2 Preanalytics —— 100
7.2.1 Specimen collection —— 100
7.2.2 Clinical circumstances for using NAT to diagnose
       HIV infection —— 101
7.2.3 Clinical circumstances for using NAT to monitor
       HIV infection —— 102
7.3 Analytics —— 103
7.3.1 Main technologies for NAT —— 103
7.3.2 HIV RNA assays —— 104
7.3.3 HIV DNA assays —— 105
7.3.4 HIV drug resistance assays —— 106
7.3.5 HIV tropism assays —— 108
7.3.6 Assays for minority HIV variants —— 109
7.4 Postanalytics —— 110

7.4.1   Molecular diagnosis of HIV infection — 110
7.4.2   Monitoring HIV infection — 110
7.5   Take-home messages — 111
7.6   Further reading — 112

8   Hepatitis viruses — 113
         Dieter Hoffmann, Thomas Michler and Ulrike Protzer
8.1   Major symptoms — 113
8.2   Preanalytics — 113
8.3   Analytics — 115
8.3.1   Adenoviruses — 116
8.3.2   HAV — 117
8.3.3   HBV — 117
8.3.4   HCV — 118
8.3.5   HDV — 119
8.3.6   HEV — 120
8.3.7   Herpes viruses — 120
8.3.8   Yellow fever virus and hemorrhagic fever viruses — 121
8.4   Postanalytics – interpretation of results — 121
8.4.1   HAV/HEV — 121
8.4.2   HBV — 121
8.4.3   HDV — 122
8.4.4   HCV — 122
8.5   Take-home messages — 123
8.6   Further reading — 123

9   Pathogens relevant in transplantation medicine — 125
         Marco Ciotti and Harald H. Kessler
9.1   Clinical manifestations — 128
9.2   Preanalytics — 128
9.2.1   Adenoviruses — 129
9.2.2   CMV — 129
9.2.3   EBV — 130
9.2.4   HHV-6 — 130
9.2.5   HHV-8 — 130
9.2.6   VZV — 131
9.2.7   BKPyV — 131
9.2.8   JCPyV — 131
9.3   Analytics — 131
9.3.1   Sample preparation — 131
9.3.2   Nucleic acids amplification and detection — 132
9.4   Postanalytics – interpretation of results — 140

9.4.1   Adenoviruses —— 140
9.4.2   CMV —— 140
9.4.3   EBV —— 141
9.4.4   HHV-6 —— 141
9.4.5   HHV-8 —— 141
9.4.6   VZV —— 141
9.4.7   BKPyV —— 142
9.4.8   JCPyV —— 142
9.5   Take-home messages —— 142
9.6   Further reading —— 143

10   **Pathogens in lower respiratory tract infections —— 145**
            Margareta Ieven and Katherine Loens
10.1   Clinical importance of different etiologic agents —— 145
10.2   Specimen collection —— 148
10.2.1   *S. pneumoniae* —— 148
10.2.2   *M. pneumoniae, L. pneumophila,* and *C. pneumoniae* —— 150
10.2.3   *B. pertussis* —— 150
10.2.4   *Respiratory viruses* —— 151
10.3   Diagnostic procedures —— 151
10.3.1   Sample preparation and nucleic acid extraction —— 151
10.3.2   Amplification and detection methods for individual agents —— 152
10.3.3   Multiplex NAATs —— 158
10.4   External quality control —— 170
10.5   The clinical usefulness and implementation of NAATs —— 171
10.6   Concluding remarks —— 172
10.7   Further reading —— 173

11   **Molecular diagnosis of gastrointestinal pathogens—— 175**
            Corinne F.L. Amar
11.1   Clinical manifestations —— 178
11.2   Preanalytics —— 178
11.3   Analytics —— 180
11.4   Postanalytics —— 186
11.4.1   Clinical sensitivity and diagnostic specificity —— 186
11.4.2   Interpretation of results —— 191
11.5   Further reading —— 192

12   **Pathogens relevant in the central nervous system —— 193**
            Helene Peigue-Lafeuille and Cécile Henquell
12.1   Clinical manifestations —— 197
12.1.1   Viral meningitis —— 197

12.1.2   Acute community-acquired bacterial meningitis —— 197
12.1.3   *Mycobacterium tuberculosis* —— 200
12.1.4   Encephalitis —— 201
12.2   Preanalytics —— 201
12.2.1   Goals of etiological investigations —— 201
12.2.2   Specimens and handling —— 201
12.2.3   Time of lumbar puncture during the course of illness
            and quantity of CSF required —— 202
12.2.4   Transport and storage of specimens —— 203
12.3   Analytics —— 204
12.3.1   Sample preparation —— 204
12.3.2   Nucleic acid amplification and detection —— 204
12.4   Postanalytics —— 208
12.4.1   Workflow and testing schedules for molecular tests —— 208
12.4.2   Limitations of molecular tests —— 210
12.4.3   Viral CNS infections —— 211
12.4.4   Bacterial CNS infections —— 211
12.4.5   Which pathogens should we look for? —— 212
12.5   Conclusion —— 213
12.6   Take-home messages —— 213
12.7   Acknowledgment —— 214
12.8   Further reading —— 214

13   **Pathogens relevant in sexually transmitted infections —— 217**
            Suzanne M. Garland and Sepehr N. Tabrizi
13.1   Symptoms and clinical manifestations —— 217
13.2   Preanalytics —— 221
13.3   Analytics —— 221
13.4   Postanalytics —— 224
13.5   Further reading —— 225

Index —— 227

Authors Index

Corinne F. L. Amar   Chapter 11
FoodBorne Pathogens Reference Unit
PHE Colindale
61 Colindale Avenue
NW9 5EQ LONDON
UK
Phone: +44(20)82004400
corinne.amar@phe.gov.uk

Stephen A. Bustin   Chapter 5
Faculty of Health, Social Care & Education
Anglia Ruskin University
Bishop Hall Lane
CHELMSFORD CM1 1SQ
UK
Phone: +44(845)1964845
stephen.bustin@anglia.ac.uk

Marco Ciotti   Chapter 9
U.O.C. di Virologia Molecolare
Fondazione Policlinico Universitatio Tor
Vergata
Viale Oxford
81-00133 ROMA
Italy
Phone: +39(6)20902087
marco.ciotti@ptvonline.it

Georg Endler   Chapter 2
Gruppenpraxis Labors.at
Praterstrasse 22
1020 WIEN
Austria
Phone: +43(1)260530
g.endler@labors.at

Suzanne M. Garland   Chapter 13
The Royal Women's Hospital
Locked Bag 300
PARKVILLE 3052
Australia
Phone: +61(3)83453671
suzanne.garland@thewomens.org.au

Dieter Hoffmann   Chapter 8
Institute of Virology
Technische Universitaet Muenchen
Trogerstrasse 30
81675 MUENCHEN
Phone: +49(89)41406825
dieter.hoffmann@virologie.med.tum.de

Margareta Ieven   Chapter 10
Vaccine & Infectious Disease Institute
University of Antwerp
Wilrijkstraat 10
2650 EDEGEM
Belgium
Phone: +32(3)8213644
greet.ieven@uza.be

Jacques Izopet   Chapter 7
Laboratoire de Virologie
Institut Federatif de Biologie
330 Avenue de Grande Bretagne, TSA 40031
31059 TOULOUSE Cedex 9
France
Phone: +33(5)67690424
izopet.j@chu-toulouse.fr

Harald H. Kessler   Chapter 4
Center for Applied Biomedicine
Medical University of Graz
Universitaetsplatz 4
8010 GRAZ
Austria
Phone: +43(316)3804363
harald.kessler@medunigraz.at

Helene Peigue-Lafeuille   Chapter 12
Laboratoire de Virologie
Centre de Biologie
CHRU Clermont-Ferrand
58, rue Montalembert
63003 CLERMONT-FERRAND
France
Phone:+33(4)73754850
hlafeuille@chu-clermontferrand.fr

Holger F. Rabenau   Chapter 1
Institute for Medical Virology
JWG-University Frankfurt/Main
Paul-Ehrlich-Strasse 40
60596 FRANKFURT/MAIN
Germany
Tel.: +49(69)63015312
rabenau@em.uni-frankfurt.de

Reinhard B. Raggam   Chapter 3
Clinical Institute of Medical and
Chemical Laboratory Diagnostics
Medical University of Graz
Auenbruggerplatz 15
8036 GRAZ
Austria
Phone: +43(316)38580243
reinhard.raggam@medunigraz.at

Ranjini Valiathan   Chapter 6
Laboratory for Clinical and
Biological Studies
University of Miami – Miller School
of Medicine
1550 NW 10th Avenue, Fox Cancer Building,
Suite 118
MIAMI, FL 33136
USA
Phone: +1(305)2432010
rvaliathan@med.miami.edu

Contributors Index

Deshratn Asthana
Laboratory for Clinical and Biological
Studies
University of Miami – Miller School
of Medicine
1550 NW 10th Avenue, Fox Cancer Building,
Suite 118
MIAMI, FL 33136
USA
Phone: +1(305)2432010
dasthan@med.miami.edu

Markus Exner
Gruppenpraxis Labors.at
Praterstrasse 22
1020 WIEN
Austria
Phone: +43(1)260530
m.exner@labors.at

Cecile Henquell
Laboratoire de Virologie
Centre de Biologie
CHRU Clermont-Ferrand
58, rue Montalembert
63003 CLERMONT-FERRAND
France
Phone: +33(4)73754850
chenquell@chu-clermontferrand.fr

Margit Hübner
Center for Applied Biomedicine
Medical University of Graz
Universitaetsplatz 4
8010 GRAZ
Austria
Phone: +43(316)3804380
margit.huebner@medunigraz.at

Eva Leitner
Center for Applied Biomedicine
Medical University of Graz
Universitaetsplatz 4
8010 GRAZ
Austria
Phone: +43(316)3804383
eva.leitner@medunigraz.at

Katherine Loens
Vaccine and Infectious Disease Institute
University of Antwerp
Wilrijkstraat 10
2650 EDEGEM
Belgium
Phone: +32(3)8202751
katherine.loens@uantwerpen.be

Thomas Michler
Institute of Virology
Technische Universitaet Muenchen
Trogerstrasse 30
81675 MUENCHEN
Phone: +49(89)41406825
thomas.michler@virologie.med.tum.de

Jamie Murphy
3rd Floor Alexandra Wing
The Royal London Hospital
LONDON E1 1BB
UK
Phone: +44(20)78828748
jamie.murphy@qmul.ac.uk

Ulrike Protzer
Institute of Virology
Technische Universitaet Muenchen
Trogerstrasse 30
81675 MUENCHEN
Phone: +49(89)41406821
protzer@virologie.med.tum.de

John Saldanha
John Saldanha Consultancy
Oakland, CA
USA
Phone: +1(510)6194713
Saldanha_ja@yahoo.co.uk

Georg Slavka
Central Laboratory
Municipal Hospital Wilhelminen
Montleartstrasse 37
1160 WIEN
Austria
Phone: +43(1)491503308
georg.slavka@wienkav.at

Sepehr N. Tabrizi
The Royal Women´s Hospital
Locked Bag 300
PARKVILLE 3052
Australia
Phone: +61(3)83453671
sepehr.tabrizi@thewomens.org.au

1 Choice of adequate sample material

Holger F. Rabenau, Reinhard B. Raggam,
Margit Hübner and Eva Leitner

Nucleic acid amplification testing (NAT) has gained major impact on the detection of pathogens. Today, NAT is widely used in the routine diagnostic laboratory. It is employed in special situations including the very early stage of infection before production of antibodies and in patients lacking antibody production due to immunosuppression. Furthermore, NAT is the method of choice to detect/exclude vertical transmission and to monitor therapy.

Reliable molecular diagnostics strongly depends on preanalytical issues including the choice of adequate sample material, optimal sampling time regarding the course of disease, and both time and conditions of the sample transport to the laboratory.

This chapter focuses on the choice of adequate sample materials for molecular diagnostics of viruses, bacteria, fungi, and protozoa. Pathogens with epidemiological and clinical significance for which molecular diagnostics plays an important role are discussed in alphabetical order.

1.1 Viruses

Adenoviruses (Family: *Adenoviridae*; approx. 50 human serotypes, subgenera A–F)
Epidemiology: worldwide distribution.
Transmission: droplets and smear infection; entrance gates are eyes and the oropharynx.
Incubation period: 5–12 days.
Clinical presentation: adenovirus infections are often asymptomatic or cause respiratory tract infections, gastroenteritis, and epidemic keratoconjunctivitis.
Complications: meningoencephalitis in children, disseminated, sepsis-like adenoviral infection with multiple organ manifestations in immunosuppressed patients.

Indication and choice of the adequate sample material for NAT:

Clinical presentation	Sample material
Epidemic keratoconjunctivitis	Conjunctival swab
Upper respiratory tract infection	Nasopharyngeal swab or aspirate, throat washing, induced sputum
Pneumonia	Bronchoalveolar lavage (BAL), EDTA whole blood
Hemorrhagic cystitis	Urine
Encephalitis	Cerebrospinal fluid (CSF)

(Continued)

(Continued)

Clinical presentation	Sample material
Gastroenteritis	Stool
Pre-emptive monitoring/suspected adenovirus infection under immunosuppression	EDTA whole blood, nasopharyngeal swab or aspirate, throat washing, urine

Astrovirus (Family: *Astroviridae*)

Epidemiology: occasional outbreaks, e.g. in nursing homes or nosocomial outbreaks in hospitals.

Transmission: smear infections or through contaminated food and water.

Incubation period: 1–3 days.

Clinical presentation: gastroenteritis with fever, vomiting and abdominal pain.

Indication and choice of the adequate sample material for NAT:

Clinical presentation	Sample material
Gastroenteritis	Stool

Bocavirus (BoV) (Family: Parvoviridae; 4 species: BoV1–BoV4)

Epidemiology: worldwide distribution, in 2–19% of patients with upper or lower respiratory tract disease predominantly during winter and spring, very common during early childhood, co-infections with other respiratory viruses frequently observed, BoV2 through BoV4 mainly in stool (enteric species), associated with gastroenteritis, co-infections with other gastrointestinal viruses in up to 100% of stool specimens.

Transmission: Transmission routes unknown; however, most likely transmitted by inhalation or contact with infectious sputum, feces, or urine.

Incubation period: Unknown.

Clinical presentations: BoV1: Respiratory tract infection with cough and wheeze, rhinorrhea, tachypnea, and fever. BoV2 through BoV4: Gastroenteritis.

Complications: Rash or exanthema, thrombopenia, pneumonia, sepsis (rarely).

Indication and choice of the adequate sample material for NAT:

Clinical presentation	Sample material
Upper respiratory tract infection	Nasopharyngeal swab or aspirate, throat washing
Pneumonia	BAL
Gastroenteritis	Stool

Note: BoV may persist in the respiratory or gastrointestinal tract as a bystander without symptoms resulting in frequent detection of BoV.

Coronaviruses (Family: *Coronaviridae*; 4 genera: alpha including the human CoVs229E and NL229E, beta including the human CoVsOC43, HKU1, MERS-CoV, and SARS-CoV, gamma including only avian pathogens, and the provisional delta genus)

Epidemiology: worldwide distribution depending on genus, high prevalence already in childhood.

Transmission: droplets and smear infection.

Incubation period: 2–5 days (SARS 2–20 days, MERS 5-12 days).

Clinical presentation: Respiratory tract infections. MERS/SARS disease with fever, cough, shortness of breath, pneumonia, and bronchiolitis; CoV-NL229E occurs especially in children with disorders of the upper respiratory tract, pneumonia, and bronchiolitis.

Indication and choice of the adequate sample material for NAT:

Clinical presentation	Sample material
Upper respiratory tract infection	Nasopharyngeal swab or aspirate, throat washing, induced sputum
Pneumonia, bronchiolitis	BAL
Suspected MERS/SARS infection	Nasopharyngeal swab or aspirate, throat washing, induced sputum, BAL

Cytomegalovirus (CMV) (Family: *Herpesviridae*)

Epidemiology: worldwide distribution, seroprevalence 50–100%.

Transmission: oro-oral contact (kissing) through saliva, and sexually through genital secretions, rarely droplets or smear infection; pre- and perinatal; possibly iatrogenic.

Incubation period: 20–60 days.

Clinical presentation: primary infection usually asymptomatic, mononucleosis-like symptoms may occur, rarely hepatitis. Re-activation is usually asymptomatic but in the immunocompromised it is potentially life threatening.

Complications: pneumonia, encephalitis, retinitis, enterocolitis and/or hepatitis; acute/chronic graft rejection.

Indication and choice of the adequate sample material for NAT:

Clinical presentation	Sample material
Pneumonia	BAL, EDTA whole blood
Encephalitis	CSF, EDTA whole blood
Disseminated CMV infection	BAL, CSF, EDTA whole blood
Colitis	Stool
Hepatitis	EDTA whole blood
Retinitis	Aqueous humor
Pre-emptive monitoring/suspected CMV infection under immunosuppression	EDTA whole blood, bone marrow, throat washing, urine, BAL, CSF, stool
Prenatal infection	Amniotic fluid, fetal EDTA whole blood
Perinatal infection	EDTA whole blood, urine

Dengue viruses (Family: *Flaviviridae*; 4 serotypes)

Epidemiology: distribution of the virus in almost all tropical and subtropical regions.

Transmission: bite through *Aedes* mosquitoes.

Incubation period: 3–7 days.
Clinical presentation: two peaked febrile illness, followed by arthralgia and rash.
Complications: re-infection can lead to dengue hemorrhagic fever (DHF) or dengue shock syndrome (DSS).

Indication and choice of the adequate sample material for NAT:

Clinical presentation	Sample material
Unclear serological constellation/serological confirmation	EDTA whole blood, serum
Suspicion or exclusion of dengue infection	EDTA whole blood, serum
DHF, DSS	EDTA whole blood, serum

Enteroviruses (Family: *Picornaviridae*; more than 100 human pathogenic types, including Polio-, Coxsackie- and ECHO viruses)
Epidemiology: worldwide distribution, infections occur mainly in summer.
Transmission: fecal-oral.
Incubation period: 2–14 (rarely up to 35) days.
Clinical presentation: frequently asymptomatic course or non-specific feverish infection ("summer flu"). Depending on the virus type, different diseases and symptoms are observed, e.g. respiratory tract infections, herpangina, acute hemorrhagic conjunctivitis, aseptic meningitis, meningoencephalitis, paralysis, eruptions, Hand-Foot-Mouth disease, myocarditis, pericarditis, hepatitis, and epidemic pleurodynia (Bornholm's disease).
Complications: severe systemic disease of newborns with meningitis, meningoencephalitis and myocarditis.

Indication and choice of the adequate sample material for NAT:

Clinical presentation	Sample material
Aseptic meningitis, meningoencephalitis	CSF, stool
Myocarditis, pericarditis, pleurodynia, hepatitis	Organ biopsy, stool
Conjunctivitis	Conjunctival swab
Respiratory tract infection, herpangina, Hand-Foot-Mouth disease	Nasopharyngeal swab or aspirate, skin swab, throat washing
Systemic disease	Stool

Note: shedding of enteroviruses in the stool can persist for months.

Epstein-Barr virus (EBV) (Family: *Herpesviridae*)
Epidemiology: worldwide distribution, seroprevalence in adults usually >90%, virus persistence with frequent (subclinical) reactivation and virus excretion.
Transmission: oro-oral contact (kissing) through saliva, and sexually through genital secretions, rarely droplets or smear infection.

Incubation period: 30–50 days.

Clinical presentation: before puberty mainly asymptomatic, afterwards infectious mononucleosis.

Complications: meningitis, encephalitis, Guillain-Barré syndrome, hepatitis, splenic rupture, hemolytic and aplastic anemia, thrombocytopenia. EBV infection is also associated with Burkitt's lymphoma, nasopharyngeal carcinoma, lymphoproliferative disease, lymphoma in immunocompromised patients (e.g. post-transplant lymphoproliferative disease [PTLD]), X-linked lymphoproliferative syndrome, oral hairy leukoplakia, and Hodgkin's lymphoma.

Indication and choice of the adequate sample material for NAT:

Clinical presentation	Sample material
Meningitis, encephalitis	CSF
Mononucleosis, aplastic anemia, Guillain-Barré syndrome	EDTA whole blood
Pre-emptive monitoring/suspected EBV infection under immunosuppression	EDTA whole blood
PTLD	EDTA whole blood
Pneumonia	BAL, EDTA whole blood
Oral hairy leukoplakia	Biopsy

Hantaviruses (Family: *Bunyaviridae*; approx. 20 hantavirus species)

Epidemiology: worldwide distribution.

Transmission: through infectious, aerosolic feces and urine of chronically infected rodents (viral reservoir).

Incubation period: 5–35 days.

Clinical presentation: hemorrhagic fever with renal syndrome (HFRS) mainly caused by Hantaan, Dobrava and Seoul species; nephropathia epidemica caused by Puumala species; Hantavirus pulmonary syndrome (HPS) caused by Sin Nombre virus.

Indication and choice of the adequate sample material for NAT:

Clinical presentation	Sample material
HFRS, nephropathia epidemica	EDTA whole blood, Urine, organ biopsy
HPS	EDTA whole blood, organ biopsy

Note: the diagnosis is usually based on serological antibody testing.

Hepatitis A virus (HAV) (Family: *Picornaviridae*)

Epidemiology: worldwide distribution, in industrialized countries the prevalence is relatively low.

Transmission: fecal-oral.

Incubation period: 3–5 weeks.

Clinical presentation: infection in childhood often asymptomatic. The risk of a symptomatic course of the disease increases with age. Unspecific symptoms such as nausea, loss of appetite, malaise, and aversion to fatty foods are followed by viral hepatitis with jaundice.

Indication and choice of the adequate sample material for NAT:

Clinical presentation	Sample material
Suspected HAV infection	EDTA plasma, serum, stool

Note: the diagnosis is usually based on serological antibody testing.

Hepatitis B virus (HBV) (Family: *Hepadnaviridae*)
Epidemiology: worldwide distribution, approximately 350 million chronic carriers.
Transmission: parenteral, vertical, sexual.
Incubation period: 1–7 months, depending on the mode of transmission and the infective dose.
Clinical presentation: often inapparent HBV infection or unspecific symptoms such as malaise, anorexia, arthralgia, and vasculitis.
Complications: tendency for development of chronic HBV infection (approximately 10%, for perinatal infection exceeding 90%) with chronic active hepatitis, liver fibrosis, liver cirrhosis, and hepatocellular carcinoma.

Indication and choice of the adequate sample material for NAT:

Clinical presentation	Sample material
Suspected acute or chronic HBV infection	EDTA plasma, serum
Therapy decision-making and monitoring	EDTA plasma, serum

Hepatitis C virus (HCV) (Family: *Flaviviridae*; at least 6 genotypes with numerous subtypes)
Epidemiology: worldwide distribution, approximately 170 million chronic carriers.
Transmission: mainly parenteral, through contaminated blood, e.g. "needle sharing" among intravenous drug abusers; rarely through needle-stick injuries in healthcare settings, sexual contact, or vertical. Infection through blood and blood products occurred frequently before introduction of antibody testing in 1991.
Incubation period: 2–26 weeks.
Clinical presentation: mainly asymptomatic or mild and unspecific symptoms with lethargy, anorexia, less than 10% are icteric.
Complications: high tendency for development of chronic HCV infection (exceeding 70%), with chronic active hepatitis, liver fibrosis, liver cirrhosis and hepatocellular carcinoma.

Indication and choice of the adequate sample material for NAT:

Clinical presentation	Sample material
Suspected acute or chronic HCV infection	EDTA plasma, serum
Therapy decision making and monitoring	EDTA plasma, serum

Note: HCV genotyping is mandatory before starting antiviral therapy.

Hepatitis D virus (HDV) (Genus: Delta virus, no family; 3 genotypes)
Epidemiology: worldwide distribution, with high prevalence in South America and low prevalence in central and northern Europe.
Transmission: mostly parenteral; sexual transmission is possible. Two possible ways of infection, either as a co-infection (HBV and HDV are transmitted simultaneously), or as a super-infection (HDV infection of an already HBV-infected individual).
Incubation period: 4 weeks to 8 months.
Clinical presentation: acute HBV-HDV co-infections often cause a severe acute hepatitis with significant mortality. The super-infection of a chronic HBV carrier with HDV usually leads to the formation of a chronic co-infection and is often more severe than HBV mono-infection.

Indication and choice of the adequate sample material for NAT:

Clinical presentation	Sample material
Suspected acute or chronic HDV infection	EDTA plasma, serum

Note: monitoring of HDV therapy may be done through monitoring of HBV load.

Hepatitis E virus (HEV) (Family: Hepeviridae, Genus: Hepevirus; 4 genotypes)
Epidemiology: most countries in the developing world (especially Southeast Asia), indigenous cases in industrialized countries. Genotypes 1 and 2 are restricted to humans, while genotypes 3 and 4 infect humans, pigs and other animal species.
Transmission: fecal-oral, through contaminated water, or food (e.g. consumption of insufficient cooked shell fish, wild boar or deer liver meat).
Incubation period: 3–8 weeks.
Clinical presentation: similar to HAV, usually self-limiting disease but chronic hepatitis E may occur, especially in immunocompromised patients. Sometimes prolonged fecal excretion. Occasionally severe liver disease (fatality rate 1–4%). Especially in pregnancy, lethality up to 25%.

Indication and choice of the adequate sample material for NAT:

Clinical presentation	Sample material
Suspected HEV infection	EDTA plasma, serum, stool

Note: Frequently false-positive serological test results; PCR testing for HEV is thus strongly recommended in cases of suspected HEV infection.

Herpes simplex virus type 1 and type 2 (HSV-1, HSV-2) (Family: *Herpesviridae*)

Epidemiology: worldwide distribution, seroprevalence in adults 75–95% (HSV-1) and 15–25% (HSV-2), virus persistence with frequent reactivation.

Transmission: mainly through oro-oral contact and through intimate contact, rarely droplets or smear infection.

Incubation period: 2–12 days.

Clinical presentation: primary HSV-1 based infection is usually asymptomatic, sometimes referred to as gingivostomatitis. Primary genital HSV-2 based infection is often associated with blistering and ulceration, pain, fever and dysuria. Reactivation of HSV-1 and HSV-2 typically leads to painful vesicular eruptions (asymptomatic reactivation with virus excretion is possible).

Complications: conjunctivitis, herpes simplex dermatitis, eczema herpeticum, generalized HSV infection with hepatitis or pneumonia, meningitis, herpes encephalitis, Bell's palsy, herpes of the neonate.

Indication and choice of the adequate sample material for NAT:

Clinical presentation	Sample material
Encephalitis, meningoencephalitis	CSF, EDTA whole blood
Retinitis	Aqueous humor
Conjunctivitis	Conjunctival swab
Bell's palsy	Saliva
Pneumonia	BAL, EDTA whole blood
Herpes of the neonate	CSF, EDTA whole blood, serum, conjunctival swab, nasopharyngeal swab or aspirate, skin swab
Generalized HSV infection	EDTA whole blood
Herpetic lesions	Swab

Note: detection of HSV DNA in saliva or mucosal swabs is possible even in clinically healthy individuals through asymptomatic viral shedding. If there is clinical evidence of herpes encephalitis, a negative HSV DNA result should not be the sole criterion for antiviral treatment discontinuation. CSF sampling should always be done before starting antiviral treatment. If vesicular eruptions are present, vesicular fluid should always be taken using swabs. Regarding suspected diagnosis of herpes of the neonate, testing for HSV DNA is also meaningful in the absence of herpetic lesions.

Human herpes virus 6 (HHV-6) (Family: *Herpesviridae*; 2 types, A and B)

Epidemiology: worldwide distribution, seroprevalence in childhood approximately 95%.

Transmission: mainly through saliva, through sexual contact or perinatal.

Incubation period: 5–15 days.

Clinical presentation: exanthema subitum (roseola infantum – 3-day fever) with fever and volatile rash.

Complications: in rare cases, encephalitis, meningitis, (fulminant) hepatitis. In immunosuppressed patients, virus reactivation possible with interstitial pneumonia and organ rejection.

Indication and choice of the adequate sample material for NAT:

Clinical presentation	Sample material
Suspected active HHV-6 infection	EDTA whole blood, serum
Interstitial pneumonia	BAL, EDTA whole blood, serum
Encephalitis, meningitis	CSF, EDTA whole blood, serum
Hepatitis (fulminant)	Liver biopsy, EDTA whole blood, serum
Pre-emptive monitoring/suspected HHV-6 infection under immunosuppression	EDTA whole blood, serum

Note: detection of HHV-6 DNA in peripheral blood lymphocytes, lymphatic tissue and biopsies is of limited clinical significance because of virus persistence.

Human herpes virus 7 (HHV-7) (Family: *Herpesviridae*)
Epidemiology: worldwide distribution.
Transmission: mainly through saliva, probably through droplets.
Incubation period: 5–15 days.
Clinical presentation: exanthema subitum (roseola infantum – 3-day fever) with fever and volatile rash.
Complications: febrile seizure, diarrhea, vomiting. In immunosuppressed patients, virus reactivation is possible with interstitial pneumonia and organ rejection.

Indication and choice of the adequate sample material for NAT:

Clinical presentation	Sample material
Suspected active HHV-7 infection	EDTA whole blood, serum
Pre-emptive monitoring/suspected HHV-7 infection under immunosuppression	EDTA whole blood, serum

Note: detection of HHV-7 DNA in peripheral blood lymphocytes, lymphatic tissue and biopsies is of limited clinical significance because of virus persistence.

Human herpes virus 8 (HHV-8) (Family: *Herpesviridae*; Kaposi's sarcoma-associated herpes virus)
Epidemiology: worldwide distribution, endemic Kaposi's sarcoma (KS) in Africa, iatrogenic (in organ transplant recipients) and HIV-associated KS. Seroprevalence in Northern Europe and the U.S. approximately 3%, with higher prevalence in risk groups (male homosexuals and bisexuals).
Transmission: mainly through sexual contact or iatrogenic.

Incubation period: a few weeks to a few months.

Clinical presentation: Kaposi's sarcoma, Castleman's disease.

Indication and choice of the adequate sample material for NAT:

Clinical presentation	Sample material
Suspected Kaposi's sarcoma	EDTA whole blood, serum, biopsy
Castleman's disease	EDTA whole blood, serum
Pre-emptive monitoring/suspected HHV-8 infection under immunosuppression	EDTA whole blood, serum

Human immunodeficiency virus type 1 (HIV-1) and 2 (HIV-2) (Family: *Retroviridae*; HIV-1 subgroup M, with subtypes A–K, subgroups O, N and circulating recombinant forms; HIV-2 subtypes A–E)

Epidemiology: worldwide distribution of HIV-1, specific risk groups include male homosexuals, intravenous drug abusers ("needle sharing"), and professional sex workers. HIV-2 can be found mainly in Western Africa and India.

Transmission: sexual contact, intravenous drug abuse, exposure to contaminated blood, and vertical (including breastfeeding).

Incubation period: 2–8 weeks until primary symptoms; 2–10 years (and longer) until AIDS is established.

Clinical presentation: primary symptoms include mononucleosis-like illness, non-specific feverish infection, sometimes maculopapular rash and acute neurological symptoms. AIDS is characterized by a cellular immune defect resulting in opportunistic infections and tumors.

Indication and choice of the adequate sample material for NAT:

Clinical presentation	Sample material
Unclear serology/borderline results of immunoblot confirmation	EDTA plasma
Suspected acute infection (serological window)	EDTA plasma
Unclear neurological symptoms, encephalopathy	CSF, EDTA plasma
Clarification of mother-to-child transmission	EDTA plasma taken from the newborn
Newborn of HIV-positive mother	EDTA plasma
Monitoring of HAART	EDTA plasma
Suspected development of drug resistance, failure of HAART	EDTA plasma

Note: if unclear serology (repeatedly reactive ELISA with negative or borderline immunoblot confirmation) exists, HIV RNA should always be tested. The newborn of an HIV-positive mother should be tested on HIV RNA immediately after birth; if HIV RNA is undetectable retesting should be done after 6–8 weeks and after 12–16 weeks. If failure of HAART is suspected, sequencing to check for antiretroviral drug resistance should be performed.

Human T-lymphotropic virus type 1 (HTLV-1) and 2 (HTLV-2) (Family: *Retroviridae*)
Epidemiology: worldwide distribution of HTLV-1, specific risk groups include intravenous drug abusers ("needle sharing"), male homosexuals, and professional sex workers. HTLV-2 is mainly found in Native Americans and South American Indians but also in Asian countries, commonly in Japan and Korea.
Transmission: sexual contact, intravenous drug abuse, exposure to contaminated blood, and vertical (including breastfeeding).
Incubation period: decades, 15–20 years for the development of adult T-cell leukemia/lymphoma (ATLL).
Clinical presentation: ATLL (with end-stage organomegaly), HTLV-associated myelopathy (HAM)/tropical spastic paraparesis (TSP) including motor and sensory changes in the extremities, spastic gait in combination with weakness of the lower limbs, clonus, and bladder dysfunction, opportunistic infections and tumors through alterations in the host's immune functions.

Indication and choice of the adequate sample material for NAT:

Clinical presentation	Sample material
ATLL, HAM, TSP, therapy monitoring	EDTA whole blood

Note: HTLV is predominantly cell-associated. Quantitation of HTLV proviral DNA in lymphocytes has high prognostic relevance.

Influenza viruses (Family: *Orthomyxoviridae*; 3 genera, A, B and C)
Epidemiology: worldwide distribution, annual epidemics in some countries and sporadic pandemics.
Transmission: droplets and smear infection.
Incubation period: 1–3 days.
Clinical presentation: systemic and respiratory disease. A sudden onset of fever, with headache, dry cough, myalgia and sore throat are observed.
Complications: pneumonia (often bacterial super-infection), myocarditis, myositis with myoglobinuria.

Indication and choice of the adequate sample material for NAT:

Clinical presentation	Sample material
Tracheobronchitis	Nasopharyngeal swab or aspirate, throat washing, BAL
Pneumonia	BAL

Measles virus (Family: *Paramyxoviridae*)
Epidemiology: worldwide distribution, significant importance in developing countries (particularly Africa) with a relatively high mortality rate.

Transmission: droplet infection.

Incubation period: 10–14 days (infectivity from approx. 5 days before the onset of rash).

Clinical presentation: typical symptoms with conjunctivitis, Koplik's spots and maculopapular (morbilliform) rash.

Complications: otitis media, diarrhea, Hecht's giant cell pneumonia, bacterial superinfections, central nervous system involvement including subacute measles encephalitis (SME), acute post-measles encephalitis (APME) and subacute sclerosing panencephalitis (SSPE).

Indication and choice of the adequate sample material for NAT:

Clinical presentation	Sample material
Suspected acute measles infection, unclear serological constellation	EDTA whole blood, nasopharyngeal swab or aspirate, oral fluid, throat washing, urine
Hecht's giant cell pneumonia	BAL
Encephalitis	CSF, brain biopsy

Note: for the diagnosis of APME and SSPE, the detection of antibodies in CSF is relevant while the use of PCR is not.

Metapneumovirus (MPV) (Family: *Paramyxoviridae*; 2 subtypes, A and B)
Epidemiology: worldwide distribution, seroprevalence in children >95%.
Transmission: droplet infection.
Incubation period: 3–7 days.
Clinical presentation: disorders of the upper respiratory tract, bronchiolitis, pneumonia.

Indication and choice of the adequate sample material for NAT:

Clinical presentation	Sample material
Upper respiratory tract infection	Nasopharyngeal swab or aspirate, throat washing, induced sputum
Pneumonia, bronchiolitis	BAL

Mumps virus (Family: *Paramyxoviridae*)
Epidemiology: worldwide distribution, increased incidence in winter and spring.
Transmission: droplet infection.
Incubation period: 18–21 days.
Clinical presentation: one- or two-sided parotitis; unilateral orchitis.
Complications: sterility, meningitis, rarely pancreatitis or diabetes mellitus.

Indication and choice of the adequate sample material for NAT:

Clinical presentation	Sample material
Suspected mumps meningitis	CSF
Parotitis	Oral fluid, urine

Norovirus (Family: *Caliciviridae;* 5 genogroups: GI, GII, and GIV relevant for humans, GIII infecting bovine species, GV infecting mice)
Epidemiology: increased incidence in winter with some tenacious nosocomial outbreaks.
Transmission: fecal-oral, aerosols, smear infection, contaminated food.
Incubation period: 10–50 h.
Clinical presentation: gastroenteritis with nausea, vomiting, diarrhea, fever, headache and myalgia.

Indication and choice of the adequate sample material for NAT:

	Sample material
Gastroenteritis	Stool, vomit

Papillomaviruses (HPV) (Family: *Papillomaviridae*; approx. 150 genotypes)
Epidemiology: worldwide distribution.
Transmission: through direct or intimate skin/mucosal contact.
Incubation period: 21–28 days.
Clinical presentation: often inapparent, infections of the skin and mucous membranes, depending on the virus type, "low risk" types (6, 11, 42, 43, 44, 54, 61, 70, 72, 81) which cause mainly benign diseases and "high-risk" types (16, 18, 31, 33, 35, 39, 45, 51, 52, 56, 58, 59, 68) which have the potential for causing malignant diseases. Diseases include: Verruca plantaris, epidermodysplasia verruciformis (EV) with skin cancer and condylomata acuminata plana, conjunctival papilloma, cervical intraepithelial neoplasia (CIN), Butchers warts, M. Bowen, cervix-, penis-, anus-, throat- and oral cavity-carcinoma.

Indication and choice of the adequate sample material for NAT:

Clinical presentation	Sample material
Condylomata acuminata, epidermodysplasia verruciformis, conjunctival papilloma, Butchers warts, CIN, HPV-associated carcinomas	Cell containing swab, biopsy

Note: pathogen detection and typing is relevant for risk assessment of cervical neoplasia.

Parainfluenza virus type 1–4 (Family: *Paramyxoviridae*; 4 serotypes)
Epidemiology: worldwide distribution, type 4 mainly found in America. In temperate latitudes, annual outbreaks occur in the winter months, mainly in children less than 3 years of age (seroprevalence in childhood approx. 90%).
Transmission: droplets infection.
Incubation period: 3–6 days.
Clinical presentation: acute respiratory tract disorders, mainly in young children. In adults, infection is usually mild or unapparent.

Indication and choice of the adequate sample material for NAT:

Clinical presentation	Sample material
Upper respiratory tract infection	Nasopharyngeal swab or aspirate, throat washing, induced sputum
Pneumonia	BAL

Parvovirus B19 (Family: *Parvoviridae*)
Epidemiology: worldwide distribution, seroprevalence in adults approximately 70%.
Transmission: droplets, possibly through blood and blood products, transplacental infection.
Incubation period: 7–10 days.
Clinical presentation: Erythema infectiosum (Fifth disease).
Complications: in immunocompetent patients, lymphadenopathy and arthralgia; in immunocompromised patients, chronic infection with chronic anemia, and thrombocytopenia. Further, meningitis, encephalopathy, myocarditis, vasculitis and hepatitis can be associated with Parvovirus B19 infection. Vertical Parvovirus B19 infection can lead to hydrops fetalis (10%), with pre-existing anemia and risk of aplastic crisis.

Indication and choice of the adequate sample material for NAT:

Clinical presentation	Sample material
Anemia	EDTA whole blood, bone marrow
Aplastic crisis	EDTA whole blood
Meningitis, encephalopathy	EDTA whole blood, CSF
Myocarditis, hepatitis, vasculitis	EDTA whole blood, biopsy
Suspected hydrops fetalis	Amniotic fluid

Note: Parvovirus B19 can persist in the bone marrow without clinical symptoms.

Polyomaviruses (BKPyV, JCPyV) (Family: *Polyomaviridae*)
Epidemiology: worldwide distribution, seroprevalence in adults up to 90%.
Transmission: probably through respiratory fluids, urine, or contaminated water.

Incubation period: unknown.

Clinical presentation: usually asymptomatic. In immunocompromised patients, BKPyV causes infection of the urinary tract with (hemorrhagic) cystitis. In severe cases, clinical manifestations may include ureter stenosis and even renal dysfunction. JCPyV can pass the blood-brain barrier and may lead to the fatal progressive multifocal leukoencephalopathy (PML) in immunosuppressed individuals, especially in AIDS patients. However, it can also cause infection of the urinary tract.

Indication and choice of the adequate sample material for NAT:

Clinical presentation	Sample material
PML	CSF, brain biopsy
Preemptive monitoring after kidney transplantation, suspected BKPyV associated nephropathy in renal-transplant recipients, hemorrhagic cystitis in allogenic bone marrow transplant recipients	EDTA whole blood, urine

Note: increased risk of systemic infection with BKPyV after kidney transplantation. If urine is tested, it has to be considered that intermittent excretion in clinically healthy individuals is also possible.

Rabies virus (Family: *Rhabdoviridae*)

Epidemiology: worldwide distribution, annually approximately 60 000 cases, mainly in developing countries.

Transmission: through a bite from, or narrow (mucosal) contact with an infected animal.

Incubation period: usually 3–12 weeks, rarely a few days up to 6 years.

Clinical presentation: in approximately 70% rabies-related encephalitis with headache, followed by tonic spasms of the pharyngeal-, laryngeal- and respiratory muscles, increased salivation, extreme hydrophobia, death (100%) as a result of heart paralysis. In approximately 30% paralytic rabies ("silent rage") similar to that of Guillain-Barré syndrome can be observed.

Indication and choice of the adequate sample material for NAT:

Clinical presentation	Sample material
Suspected rabies	Skin biopsy
Suspected rabies encephalitis	CSF, brain tissue

Respiratory syncytial virus (RSV) (Family: *Paramyxoviridae*; 2 types, A and B)

Epidemiology: worldwide distribution, epidemic infections in late autumn and winter.

Transmission: droplets and smear infection.

Incubation period: 3–7 days.

Clinical presentation: fever with rhinitis, laryngitis, bronchiolitis and pneumonia. For infants <4 months, RSV infection is partially life threatening. Older children and adults mainly have milder symptoms such as rhinitis and tracheobronchitis.
Complications: otitis media, apnea.

Indication and choice of the adequate sample material for NAT:

Clinical presentation	Sample material
Rhinitis, laryngitis, tracheobronchitis, bronchiolitis, pneumonia	Nasopharyngeal swab or aspirate, BAL
Otitis media	Nasopharyngeal swab or aspirate

Rhinovirus (Family: *Picornaviridae*; more than 100 serotypes; 3 species: RV-A, RV-B, RV-C)
Epidemiology: RV causes respiratory illnesses, including the so called "common cold". Distribution is worldwide and affects all age groups. Prevalence is throughout the year with peaks in early fall and spring.
Transmission: Droplets and smear infection.
Incubation period: 12 hours to 3 days.
Clinical presentations: Rhinitis, rhinosinusitis, pharyngitis, acute otitis media, bronchiolitis, pneumonia.
Complications: Asthma exacerbation, acute exacerbation of chronic obstructive pulmonary disease and cystic fibrosis, fatal pneumonia in immunocompromised patients after solid organ/bone marrow transplantation and in hematopoetic stem cell transplant recipients.

Indication and choice of the adequate sample material for NAT:

Clinical presentation	Sample material
Respiratory tract infection, pneumonia	Nasopharyngeal swab or aspirate, throat washing, induced sputum, BAL
Otitis media	Nasopharyngeal swab or aspirate

Notes: Rhinovirus shedding normally stops within 11 to 21 days. Persistence may represent serial infections.

Rotavirus (Family: *Reoviridae*, serogroups A–G)
Epidemiology: worldwide distribution, in developing countries every year more than 450 000 rotavirus-related deaths in children.
Transmission: fecal-oral.
Incubation period: 1–3 days.

Clinical presentation: severe gastroenteritis with diarrhea and vomiting, fever and dehydration, especially during infancy.
Complications: encephalitis, exsiccosis.

Indication and choice of the adequate sample material for NAT:

Clinical presentation	Sample material
Gastroenteritis	Stool
Suspected rotavirus encephalitis	CSF

Rubella virus (Family: *Togaviridae*)
Epidemiology: worldwide distribution.
Transmission: droplets, transplacental infection.
Incubation period: 10–21 days.
Clinical presentation: flu-like symptoms with nuchal lymphadenopathy and small dotted exanthema; in adults, partially volatile arthritis.
Complications: vertical rubella infection in the first trimester of pregnancy can lead to embryopathy with spontaneous stillbirth (20%) or to the congenital rubella syndrome with sensorineural deafness, eye abnormalities (especially retinopathy, cataract, and microphthalmia), and congenital heart disease (especially patent ductus arteriosus) and sometimes even mental retardation.

Indication and choice of the adequate sample material for NAT:

Clinical presentation	Sample material
Suspected vertical rubella infection	EDTA whole blood, chorionic villi biopsy, amniotic fluid, fetal blood
Suspected congenital rubella infection	Urine, nasopharyngeal swab or aspirate, lens material, EDTA whole blood, CSF
Cataract of the newborn	Aqueous humor, lens material
Suspected acute rubella infection, unclear serological constellation	EDTA whole blood, oral fluid, nasopharyngeal swab or aspirate, throat washing

Note: if vertical transmission of rubella infection is suspected, a chorionic villi biopsy can be used during weeks 11–18 of pregnancy for NAT testing (besides maternal EDTA whole blood testing). Amniotic fluid should be tested during weeks 18–22 of pregnancy; afterwards IgM antibody testing of fetal blood is recommended.

Tick-borne encephalitis virus (TBEV) (Family: *Flaviviridae*; 3 subtypes)
Epidemiology: TBEVs are endemic to forest areas in the majority of European countries. TBE is the most important arthropod-transmitted viral disease in Europe.
Transmission: tick-bite.
Incubation period: 7–21 days.

Clinical presentation: frequently biphasic illness, fever prior to neurological symptoms. Of those infected, 10–30% develop a severe neurological disease, such as meningitis, meningoencephalitis, or meningoencephalomyelitis.
Complications: long-term sequelae (e.g. flaccid paralysis); the case-fatality rate is up to 5%.

Indication and choice of the adequate sample material for NAT:

Clinical presentation	Sample material
Neurological symptoms	CSF

Note: the detection of specific antibodies in serum and in CSF is currently the diagnostic method of choice.

Varicella zoster virus (VZV) (Family: *Herpesviridae*)
Epidemiology: worldwide distribution, seroprevalence in adults more than 95%.
Transmission: droplets, mucous membrane contact, transplacental infection.
Incubation period: 10–23 days.
Clinical presentation: primary infection is also referred to as chickenpox. During childhood, it usually shows a milder course; in adults, especially in immunocompromised patients, secondary bacterial infections, sepsis, hemorrhagic chickenpox, pneumonia, or encephalitis can occur. VZV reactivation leads to herpes zoster (shingles).
Complications: post-zoster neuralgia, secondary bacterial infections, encephalitis, generalized zoster infection with septicemia, zoster ophthalmicus, Ramsey-Hunt syndrome. In cases of primary infection between week 13 and week 20 of pregnancy the varicella congenital syndrome can be observed (in approx. 2%) with skin lesions in dermatomal distribution, neurologic defects, eye diseases and skeletal anomalies. A perinatal infection results in chickenpox of the neonate with significant lethality up to 35%.

Indication and choice of the adequate sample material for NAT:

Clinical presentation	Sample material
Encephalitis	CSF, EDTA whole blood
Pneumonia	BAL, EDTA whole blood
Hemorrhagic chickenpox	Skin swab, skin biopsy, EDTA whole blood
Ramsey-Hunt syndrome	Oral fluid, skin swab
Chickenpox of the neonate	Skin swab, EDTA whole blood
Clarification of a florid varicella-zoster virus infection	Skin swab
Generalized varicella zoster infection	Skin swab, EDTA whole blood

Note: sampling should always be done before starting antiviral treatment. If vesicular eruptions are present, vesicular fluid should always be taken using swabs for sampling. If oral fluid is tested, it has to be considered that viral shedding into the oral cavity can also be observed in healthy individuals.

West Nile virus (WNV) (Family: *Flaviviridae*)
Epidemiology: distribution in Africa, parts of Europe, India, Israel, USA.
Transmission: bite through *Culex* spp. mosquitoes. Vertical transmission possible.
Incubation period: 3–14 days.
Clinical presentation: mostly asymptomatic course (80%), fever with flu-like symptoms, headache and eruptions.
Complications: encephalitis, menigoencephalitis, flaccid paralysis (especially in persons aged over 70 years).

Indication and choice of the adequate sample material for NAT:

Clinical presentation	Sample material
Encephalitis, meningoencephalitis, flaccid paralysis	CSF, EDTA whole blood
Suspected West Nile virus infection	EDTA whole blood

1.2 Bacteria

Bordetella spp. (Family: *Proteobacteria*; aerob; 3 species, *Bordetella pertussis, Bordetella parapertussis, Bordetella bronchiseptica*)
Epidemiology: worldwide distribution with seasonal increases in wintertime.
Transmission: droplets.
Incubation period: 7–20 days.
Clinical presentation: whooping cough, starting with an initial catarrhal phase with symptoms similar to those of the common cold, proceeding with a dry and persistent cough.
Complications: subconjunctival bleedings, otitis media, pneumonia, apnea.

Indication and choice of the adequate sample material for NAT:

Clinical presentation	Sample material
Whooping cough	Nasopharyngeal swab or aspirate

Note: the combination of real-time PCR and single-serum serology (IgA) are currently the most efficient diagnostic tools. The molecular assay should be able to distinguish between *B. pertussis* and *B. parapertussis*.

Borrelia burgdorferi (Family: *Spirochaetaceae*)
Epidemiology: most common species in Europe and USA, mostly in the warm months of the year. In endemic areas, 2–50% of the ticks are infected.
Transmission: tick-bite.
Incubation period: depending on the stadium of the disease, stadium I: 1–16 weeks, stadium II: months, stadium III: years.

Clinical presentation: stadium I: erythema chronicum migrans (ECM); Stadium II: facial palsy, meningoencephalitis, myocarditis, lymphadenosis cutis benigna; Stadium III: arthritis, acrodermatitis chronica atrophicans, polyneuropathia; neuroborreliosis.

Indication and choice of the adequate sample material for NAT:

Clinical presentation	Sample material
Arthritis	Joint puncture fluid
Neuroborreliosis	CSF

Note: high analytic sensitivity required because of low bacterial concentration.

Chlamydia trachomatis (Family: *Chlamydiaceae*)

Epidemiology: worldwide distribution, genital chlamydia infection is the most frequently diagnosed bacterial sexually transmitted infection worldwide, infection does not prevent against re-infection.

Transmission: Chlamydia trachomatis is transmitted during unprotected vaginal, anal or oral sex and can be passed from an infected mother to the newborn during vaginal delivery.

Incubation period: 10–24 days.

Clinical presentation: urethritis, cystitis, cervicitis, lymphogranuloma venerum, inclusion-body conjunctivitis.

Complications: pelvic inflammatory disease, prostatitis, epididymitis, sterility, arthritis (Mb. Reiter), perinatal infection with trachoma and blindness of the newborn.

Indication and choice of the adequate sample material for NAT:

Clinical presentation	Sample material
Cervicitis	Cervical swab (cell containing)
Urinary tract infections, prostatitis, epididymitis, arthritis dysenterica (Reiter's syndrome)	Urethral swab (males), urine

Chlamydophila pneumoniae (Family: *Chlamydiaceae*)

Epidemiology: worldwide distribution, *Chlamydophila pneumoniae* affects all age groups and is most common among the older teenage-age and the 60–79 year-old groups. Re-infection is common after a short period of immunity. The incidence is 1/1000 per year. *Chlamydophila pneumoniae* causes 10% of community-acquired pneumonias.

Transmission: droplets.

Incubation period: 7–28 days.

Clinical presentation: respiratory tract infections including pharyngitis, laryngitis, pneumonia.

Indication and choice of the adequate sample material for NAT:

Clinical presentation	Sample material
Respiratory tract infections, pneumonia	Induced sputum, BAL

Clostridium difficile (Family: *Clostridiaceae*, anaerob)

Epidemiology: distribution in industrialized countries, responsible for both hospital-acquired and community-acquired diarrhea leading to a major public health problem.
Transmission: fecal-oral route.
Risk factors: healthcare environment, antimicrobial treatment, age ≥65, immunosuppression, chronic underlying disease, proton pump inhibitors.
Clinical presentation: diarrhea.
Complications: life threatening bowel complications.

Indication and choice of the adequate sample material for NAT:

Clinical presentation	Sample material
Diarrhea	Unformed stool

Note: Only *Clostridium difficile* producing toxins (A/B) are leading to infection; therefore, the NAT should target toxin genes.

Enterococcus, **vancomycin-resistant (VRE)** (Family: *Enterococcaceae, aerob*)

Epidemiology: Enterococcus faecalis (90–95%) and *Enterococcus faecium* (5–10%) are common commensal microorganisms in the intestines of humans: the most important feature of this genus is the high level of endemic antibiotic resistance. In the last two decades, particularly virulent strains of *Enterococcus* that are resistant to vancomycin (VRE) have emerged in nosocomial infections of hospitalized patients.
Clinical presentation: important clinical infections caused by enterococci include wound infections, urinary tract infections, bacteremia, bacterial endocarditis, diverticulitis with diarrhea and meningitis.
Complications: sepsis with an overall rate of lethality up to 58%.

Indication and choice of the adequate sample material for NAT:

Clinical presentation	Sample material
Wound infections	Swab
Meningitis	CSF
Endocarditis, septicemia	EDTA whole blood

Note: the phenotypes VanA and VanB are the most common acquired VREs. For infection control purposes, the identification of species level is mandatory to distinguish from non-transferable intrinsic VanC resistance.

Helicobacter pylori (Family: *Helicobacteriaceae*, microaerophilic)
Epidemiology: worldwide distribution. Helicobacter *pylori* infects over 50% of the world's population but only a small subset of infected people develop *Helicobacter pylori*-associated disease.
Transmission: oral-oral, gastro-oral or fecal-oral route.
Clinical presentation: abdominal pain (stomach ache), nausea.
Complications: chronic gastritis, peptic ulcer disease, mucosa-associated lymphoid tissue (MALT) lymphoma, gastric carcinoma.

Indication and choice of the adequate sample material for NAT:

Clinical presentation	Sample material
Gastritis	Biopsy

Legionella pneumophila (Family: *Legionellaceae*, aerob)
Epidemiology: worldwide distribution, common sources of Legionella include cooling towers, air conditioning systems, domestic hot water systems, fountains and similar disseminators that draw upon a public water supply; natural sources include freshwater ponds and creeks.
Transmission: aerosols.
Incubation period: 2–10 days.
Clinical presentation: legionellosis with multifocal necrotizing pneumonia; pontiac fever (without pneumonia).
Complications: fatality rate exceeding 20%.

Indication and choice of the adequate sample material for NAT:

Clinical presentation	Sample material
Legionellosis, pneumonia	BAL

***Mycobacterium tuberculosis* complex** (Family: *Mycobacteriaceae*, aerob)
Epidemiology: worldwide distribution, approximately 15 million infections/year, classified into *M. tuberculosis, M. bovis, M. africanum, M. microti, M. canetti, M. pinnepedi, M. caprae, M. mungi,* and the vaccination strain *Bacillus Calmette-Guerin (BCG).*
Transmission: mainly droplets.
Incubation period: 4–12 weeks, sometimes years.
Clinical presentation: unapparent infection, fever, night sweats, and weight loss, cough.
Complications: pneumonia, pleuritis, miliary tuberculosis, meningitis, arthritis, osteomyelitis, urogenital tuberculosis, sepsis landouzy (rare).

Indication and choice of the adequate sample material for NAT:

Clinical presentation	Sample material
Pulmonary tuberculosis	Induced sputum, BAL, pleural effusion
Extrapulmonary tuberculosis	CSF, urine, bone marrow, organ tissues, lymphatic tissue, joint puncture fluid

Mycoplasma pneumoniae (Family: *Mycoplasmataceae*, aerob)
Epidemiology: worldwide distribution.
Transmission: droplets.
Incubation period: 7–28 days.
Clinical presentation: respiratory tract infections including pharyngitis, tracheobronchitis, interstitial pneumonia, bronchiolitis.
Complications: subsegmental and segmental atelectasis of the lung; otitis media.

Indication and choice of the adequate sample material for NAT:

Clinical presentation	Sample material
Respiratory tract infections, pneumonia	Induced sputum, BAL

Note: NAT enables epidemiological monitoring of macrolide resistance to *M. pneumoniae*.

***Neisseria* species** (Family: *Neisseriaceae*, aerob; 2 human pathogenic species, *Neisseria meningitidis* and *Neisseria gonorrhoeae*)
Epidemiology: worldwide distribution.
Transmission: droplets *(N. meningitidis)*; sexually transmitted or perinatal *(N. gonorrhoeae)*.
Incubation period: 2–10 days *(N. meningitidis)*; 2–7 days, sometimes weeks *(N. gonorrhoeae)*.
Clinical presentation: for *N. meningitidis*: meningitis, sepsis; for *N. gonorrhoeae*: gonorrhea with purulent (or pus-like) discharge from the genitals which may be foul smelling, inflammation, redness and swelling of the outer genitals, acute urethritis.
Complications: for *N. gonorrhoeae*: pelvic inflammatory disease, bartholinitis, sterility (women); prostatitis, proctitis, epididymitis, sterility (men); disseminated gonococcal infection with sepsis, endocarditis, meningitis; perinatal infection with conjunctivitis purulenta.

Indication and choice of the adequate sample material for NAT:

Clinical presentation	Sample material
Meningitis	CSF
Gonorrhea, urethitis, prostatitis	Urethral swab, urine (men)
Cervicitis	Cervical swab
Pelvic inflammatory disease, bartholinitis, proctitis	Vaginal/cervical/anal swab
Conjunctivitis	Conjunctival swab
Sepsis	EDTA whole blood, CSF

Methicillin-resistant *Staphylococcus aureus* (MRSA) (Family: *Staphylococcaceae*, aerob)

Epidemiology: worldwide distribution. MRSA is sub-categorized as community-acquired MRSA (CA-MRSA) or healthcare-associated MRSA (HA-MRSA).

Transmission: smear infection.

Incubation period: highly variable.

Clinical presentation: wound infections, pneumonia.

Complications: necrotizing pneumonia (especially CA-MRSA), endocarditis and sepsis.

Indication and choice of the adequate sample material for NAT:

Clinical presentation	Sample material
Wound infection	Swab
Pneumonia	BAL
Bacteremia, endocarditis	EDTA whole blood
Screening	Nasal swab

Note: detection is mainly based on the mecA gene. Blood culture is still mandatory for the diagnosis of bacteremia.

Group B *Streptococcus* (Family: *Streptococcaceae*, aerob)

Epidemiology: worldwide distribution, Group B *Streptococcus* is part of the normal flora of the gut and genital tract and is found in 20–40% of women, carriers of the organism are asymptomatic.

Transmission: perinatal infection.

Incubation period: a few hours.

Clinical presentation: neonatal infections including pneumonia, meningitis, septicemia.

Complications: sepsis, hearing loss as long-term sequelae of Group B *Streptococcus*-meningitis.

Indication and choice of the adequate sample material for NAT:

Clinical presentation	Sample material
Suspected Group B *Streptococcus* infection	Vaginal or cervical swab, anal swab
Meningitis	CSF
Bacteremia, sepsis	EDTA whole blood

Note: NAT for rapid intrapartum screening.

Streptococcus pneumoniae (Family: *Streptococcaceae*, aerob)
Epidemiology: worldwide distribution, *Streptococcus pneumoniae* can be part of the normal upper respiratory tract flora (5–10% of healthy adults and 20–40% of healthy children) with the potential to become pathogenic.
Transmission: droplets, endogenous infection.
Incubation period: highly variable.
Clinical presentation: respiratory tract infections, sinusitis, pneumonia, otitis media, meningitis.
Complications: sepsis, brain abscess.

Indication and choice of the adequate sample material for NAT:

Clinical presentation	Sample material
Respiratory tract infections, pneumonia	Nasopharyngeal swab, induced sputum, BAL, pleural effusion
Meningitis	CSF
Bacteremia, sepsis	EDTA whole blood

Note: blood culture is still mandatory for the diagnosis of sepsis.

1.3 Fungi

Aspergillus species (Group: Molds; more than 100 species)
Epidemiology: worldwide distribution. Approximately 10 species are medically relevant including *Aspergillus fumigatus*, *Aspergillus niger*, *Aspergillus terreus*, and *Aspergillus flavus*.
Transmission: spore inhalation.

Clinical presentation: pulmonary aspergillosis includes allergenic bronchopulmonal aspergillosis (ABPA), aspergilloma in the lung and pneumonia (immunocompromised patients). Extrapulmonary aspergillosis includes rhino sinusitis, allergic fungal sinusitis (AFS), aspergilloma in the brain and endocarditis.

Complications: invasive aspergillosis, sepsis.

Indication and choice of the adequate sample material for NAT:

Clinical presentation	Sample material
ABPA	BAL, lung tissue
Aspergilloma, AFS, encephalitis, endocarditis	Fungus ball, sinunasal mucus, CSF, organ tissue
Acute invasive aspergillosis, sepsis	EDTA whole blood, CSF

Note: detection of aspergillus DNA in EDTA whole blood samples is not necessarily associated with invasive aspergillosis. In addition to serum aspergillus antigen (galactomannan) testing for monitoring, molecular testing can enhance diagnostic sensitivity in patients at risk.

Candida **species** (Group: Yeasts; more than 150 species)

Epidemiology: worldwide distribution. Approximately 7 species are medically relevant including *Candida albicans*, the most significant pathogenic species, *Candida tropicalis, Candida glabrata, Candida krusei, Candida parapsilosis, Candida dubliniensis and Candida lusitaniae.*

Transmission: endogenous infection.

Clinical presentation: in immunocompetent persons, candidiasis usually presents as a localized infection of the skin or mucosal membranes, including the oral cavity (thrush), the pharynx or esophagus, the gastrointestinal tract, the urinary bladder, or the genitalia, causing vaginal irritation, vaginitis or balanitis.

Complications: in immunocompromised patients *Candida* spp. has the potential to become systemic, causing candidemia and sepsis.

Indication and choice of the adequate sample material for NAT:

Clinical presentation	Sample material
Mucocutaneous candidiasis	Oral swab, vaginal swab
Pneumonia	BAL, lung tissue, bronchial/tracheal secretion
Invasive candidiasis	EDTA whole blood

Note: the diagnosis of pulmonary candidiasis is based on histological demonstration of the yeast in lung tissue with associated inflammation. Early detection of candida DNA in whole blood samples enables earlier commencement of antifungal therapy. However, the use of beta-D-glucan testing in serum may be superior for the diagnosis and therapy monitoring of candidemia.

Pneumocystis jirovecii (former *carinii*) (Family: *Pneumocystidaceae*)
Epidemiology: worldwide distribution. *Pneumocystis jirovecii* can be found in lungs of healthy people and as a source of opportunistic infection.
Transmission: airborne route, endogenous infection.
Clinical presentation: pneumonia
Complications: life-threatening pneumonia in immunocompromised patients (e.g. AIDS patients).

Indication and choice of the adequate sample material for NAT:

Clinical presentation	Sample material
Pneumonia	BAL, lung tissue

1.4 Protozoa

Toxoplasma gondii (Family: *Sarcocystidae*)
Epidemiology: worldwide distribution, estimated seroprevalence between 30% and 65%, with large variations between countries. Primary host is the felid (cat) family.
Transmission: ingestion of raw or partly cooked meat containing *Toxoplasma* cysts, hand-to-mouth contact after handling undercooked meat or contaminated cat feces, contaminated drinking water, transplacental infection or receiving an infected organ transplant or blood transfusion.
Incubation period: 1–2 weeks.
Clinical presentation: mostly asymptomatic with mild fever and swollen lymph nodes.
Complications: in immunocompromised patients and following congenital infection, severe toxoplasmosis with encephalitis and necrotizing retinochoroiditis.

Indication and choice of the adequate sample material for NAT:

Clinical presentation	Sample material
Suspected infection	CSF, amniotic fluid, EDTA whole blood, aqueous humor

2 Stability of the specimen during preanalytics

Georg Endler, Georg Slavka and Markus Exner

Although sample processing is usually well standardized and covered by established standards in the molecular diagnostic laboratory, preanalytical procedures outside the laboratory usually follow unwritten traditions with considerable inter-individual variability. Studies have shown that preanalytical errors make up to 85% of all laboratory errors, 95% of them occurring outside the laboratory. In particular, molecular assays are sensitive to suboptimal preanalytical conditions. False-negative results may occur due to degradation of nucleic acids during transport or polymerase chain reaction (PCR) inhibitors, resulting in delayed or misdiagnosis of infections. However, contamination may cause false-positive results, which could have severe consequences. The main issues of concern during sample transport and storage include nucleic acid integrity, contamination, sample identity, and the risk of environmental hazards due to infectious material.

2.1 Sample integrity during collection

Prior to sample collection standardized procedures might help to reduce frequent pre-analytical errors in clinical practice. Standard procedures to reduce frequent errors do not differ significantly from standard sample collection procedures and have been discussed extensively in various consensus documents. These include correct patient identification, infection control, and exclusive use of single use disposals (including needles and collection devices). Since sample collection for these rather specialized tests is usually not done routinely by the staff, it is advisable to provide a short one page summary to facilitate sample collection including the following steps:
1. Selection of appropriate sample collection devices
2. Correct identification of the patient (in up to 2% of all samples, wrong patient identification can be suspected) and appropriate labeling of the collection tubes
3. Sample collection
4. Safe disposal of needles and other potentially infectious items
5. Storage and shipment procedures

The following general sampling considerations for selected materials may be helpful.

2.1.1 Blood

Blood sampling for viral nucleic acid testing does not differ significantly from standard phlebotomies. Due to DNase inhibition in EDTA, viral nucleic acids are more stable in EDTA whole blood/plasma; however, native blood or citrate tubes may also be used. To prevent cross contamination within the laboratory by multipipettes, it is highly advisable to use separate tubes. Genomic nucleic acids are generally relatively stable in blood. Storage and transport at room temperature is usually sufficient for up to a 6-hours period. If storage or transport requires a longer period, samples should be stored at 4°C, preferably no longer than 3 days.

2.1.2 Urine

In general, urine samples should be collected in sterile tubes without additives. Since fragile bacteria may degrade easily in urine, the sample should be stored at 4°C if transport to the laboratory is not possible within 2 h. Yield of pathogens (especially chlamydia and neisseria) is usually higher, if the first morning urine is sampled. Midstream samples of urine should not be used for detection of genital infections due to the low copy number of pathogens.

2.1.3 Stool

Regarding stability and storage conditions for stool samples, no validated studies have been published so far. Hence, the following standard recommendation should be applied: Stool samples should be kept at 4°C if transport to the laboratory takes more than 6 h. Yet, experience shows that stool samples are usually stable even if kept at room temperature for up to 48 h.

2.2 Degradation of DNA

In biological matrices, DNA may be degraded rapidly due to the presence of desoxy-ribonucleases (DNases). *In vivo*, DNases play a major role to ensure individual genomic integrity. Thus, it is not surprising that DNases also represent a major defense mechanism against biological pathogens in the induction of apoptosis. Apoptosis is characterized by cell shrinkage, nuclear condensation, and internucleosomal DNA cleavage. Besides the central role of caspases and other proteases, cell death triggers DNA degradation so that DNases have an active role in apoptotic cell death. The best-characterized apoptotic DNase is CAD, a neutral Mg-dependent endonuclease. Its activity is regulated by its inhibitor, ICAD which is cleaved by caspases. Other neutral DNases such as endonuclease G and GADD have been shown to cleave nuclear DNA in apoptotic conditions. In cells, the cytosolic pH is maintained at 7.2, mostly due to the activity of the Na^+/H^+ exchanger. In many apoptotic conditions, a decrease in the intracellular pH has been shown. This decrease may activate different acid DNases,

Tab. 2.1: Strategies for inactivation of DNases.

Depletion of Mg^{2+} and Ca^{2+} ions through chelating agents (e.g. EDTA)
Adjustment of pH to 7.5
Storage at −20°C or lower
Use of dried specimens

mostly when the pH decreases below 6.5. Three acidic DNases II are known at present: DNase IIa, DNase IIβ and L-DNase II. Apart from dedicated DNases, several proteins also have DNase activity including, for example lactoferrin. Thus, DNases are present in all body fluids fulfilling a variety of physiological functions. In patients with cystic fibrosis, inhalatory DNases have been successfully applied to reduce the viscoelasticity of sputum and to enhance the clearance of secretions.

Although essential *in vivo*, DNases are the main cause of *in vitro* DNA degradation in the biologic matrix, which can lead to a false-negative result. Because all DNases are either Mg^{2+} or Ca^{2+} dependent and usually activated at a pH of 6.5 or lower, inactivation of DNases can be achieved easily through adequate strategies (Tab. 2.1). In daily laboratory practice, it is advisable to store native specimens intended for DNA diagnostics at −20°C or lower for long-term storage. In contrast, dried specimens can be stored at ambient temperature.

Recent studies showed that viral and bacterial DNA seems far more stable in EDTA anticoagulated blood when compared to serum samples at room temperature. While the viral load considerably decreased in serum within 24 h at room temperature (probably due to the activity of endogenous nucleases), it remained remarkably stable for at least 3 days in EDTA anticoagulated blood.

Purified DNA can be stored safely in Tris–EDTA buffer at room temperature for 6 months or at 2–8°C for at least 1 year in the absence of DNases. Storage may be extended to up to 7 years at −20°C and to a longer period at −70°C or lower. Samples of questionable purity should always be stored at or below −20°C to ensure DNA integrity. The freezers used to store purified DNA should not be the 'frost-free' type because this type of freezer cycles temperatures continually which may lead to deterioration of nucleic acids by shearing.

2.3 Degradation of RNA

Ribonuclease (RNase) is a type of nuclease that catalyzes the degradation of RNA into smaller components. All organisms studied contain several RNases of different classes showing that RNA degradation is a very ancient and important process. Besides cleaning of cellular RNA that is no longer required, RNases play a key role in the maturation of RNA molecules including messenger RNAs and non-coding RNAs.

In addition, active RNA degradation systems are a first defense against RNA viruses and provide the underlying machinery for more advanced cellular immune strategies such as RNAi. Some cells also secrete large quantities of non-specific RNases. RNases are thus extremely common, resulting in a very short lifespan for any RNA that is not in a protected environment. It is worth noting that all intracellular RNAs are protected from RNase activity by a number of strategies including 5′ end capping, 3′ end poly-adenylation, and folding within an RNA protein complex (ribonucleoprotein particle or RNP).

RNA analysis and quantification require specimens with completely intact RNA to produce optimal results. Although nonenzymatic hydrolysis of phosphodiester bonds is favored by high temperature or pH and the presence of divalent cations (Mg^{2+}, Mn^{2+}), an RNA sample is most likely to be degraded rapidly by a contaminating RNase. Complete removal or inactivation of RNases during RNA extraction procedures has proven to be very difficult. In fact, RNases may even be introduced into the sample accidentally during handling. There are several possible sources for introduction of RNases in the laboratory. RNases are ubiquitous in the environment and are found in pollen, dust, and fingerprint grease.

If no specific RNase inhibitors are used, long-term storage of diagnostic RNA samples should be done preferably at −80°C or lower to inhibit RNase activity, as RNase may limit the stability of RNA even in frozen samples at −20°C. RNase inhibitors such as concentrated formamide or 4M-guanidine isothiocyanate have been applied successfully for long-term storage of RNA up to 18 months; however, they may interact with downstream applications such as RNA isolation or cDNA synthesis. Recently, ready-to-use RNA stabilization solutions have been brought on the market allowing storage of diagnostic RNA samples at 37°C for 1 day, 21°C for 1 week, 4°C for 1 month, and −20°C for long-term storage.

2.4 Inhibitors of PCR

For as long as the technique of PCR has been used, inhibition has been an obstacle to success. All who use PCR are likely to be impacted by inhibitors at some time with the wide range of non-blood specimens used for detection of pathogens and the often suboptimal sampling conditions making PCR based assays for pathogen detection in non-blood specimens especially vulnerable. Inhibitors generally exert their effects through direct interaction with DNA or interference with DNA polymerases. Direct binding of agents to single-stranded or double-stranded DNA may prevent amplification and facilitate co-purification of the inhibitor and the DNA. Inhibitors may also interact directly with the DNA polymerase to block enzyme activity. Furthermore, the cofactor requirements of DNA polymerases may be the target of inhibition. Magnesium is a critical cofactor and agents that reduce Mg^{2+} availability or interfere with binding of Mg^{2+} to the DNA polymerase can inhibit PCR. The presence of

Tab. 2.2: Specimen types and inhibitors.

Type of specimen	Possible inhibitors
Blood	Heme, hemoglobin, immunoglobulin G, lactoferrin, heparin
Urine	Urea
Respiratory	Acidic polysaccharides
Feces	Bile salts, complex polysaccharides
Tissue	Collagen, myoglobin

inhibitors in specimens has been described in many publications. Common specimen types known to contain inhibitors include blood, sputum, urine, feces, and tissues (Tab. 2.2). Additional sources of inhibitors may be materials and reagents that are exposed to samples during processing or DNA purification. These include excess potassium chloride, sodium chloride and other salts, ionic detergents such as sodium deoxycholate, sarkosyl and sodium dodecyl-sulfate, ethanol and isopropanol, and phenol.

The best way to avoid PCR inhibition is to prevent the inhibitor from being processed with the sample. For inhibitors that are inherent to the sample, as is the case for blood and certain body fluids, this is not always possible. In particular, vessels containing heparin should be avoided. Heparin represents one of the most potent PCR inhibitors being usually not removed completely through standard nucleic acid purification protocols. When a nucleic acid purification protocol is introduced, the method's ability to obtain efficiently inhibitor-free DNA from a wide range of sample types should be evaluated. Furthermore, techniques proven to eliminate inhibitors from the purified template DNA should be favored. There are several options to avoid effects from inhibitors not eliminated during extraction. Firstly, the choice of the DNA polymerase may have a significant impact on avoidance of inhibition. For instance, AmpliTaq® DNA polymerase which is the standard for use with several commercial PCR kits is known to be among those most sensitive to inhibition. This underscores the importance of sample handling and extraction. Secondly, increasing the amount of DNA polymerase or using additives such as bovine serum albumin which provides some resistance to inhibitors in blood are proven methods. Finally, adding less DNA template to the amplification can often improve performance significantly, if the assay sensitivity is sufficient to obtain valid results from templates that contain inhibitors.

2.5 How can contamination during specimen collection and in the laboratory be avoided?

During specimen collection, standard precautions as used for standard microbiology tests are usually sufficient to minimize the risk of contamination. However, it is essential that all vessels used are suitable for molecular tests and are free of nucleic acids and nucleases.

Due to the extreme sensitivity of PCR, sample contamination in the laboratory resulting in false-positive results is another important issue. Contamination of the PCR reaction mixture may happen through previously amplified sequences or nucleic acids from the PCR operator or other positive samples. It is therefore indispensable that the sample remains tightly capped and not used for any other test prior to molecular testing. Especially, pipettors in clinical chemistry analyzers have been reported to be a possible source of cross contamination. Therefore, it would be preferable to perform molecular testing on blood samples out of unopened primary blood tubes.

2.6 How can the sample identity be ensured?

Most errors in sample identification occur at the time of specimen collection. Thus, it is highly recommended that the laboratory provides well-defined standard operating procedures to minimize errors due to wrong patient identification.

The patient and the patient's specimen must be clearly identified at the time of collection. Specimens for molecular testing must be identified with a firmly attached label bearing at least an identification number, the patient's full name and date of birth (the patient's name alone is not sufficient for uniquely identifying a patient), and the specimen source (i.e. the type of tissue from which the specimen was taken). As suggested by the NCCLS, each sample should be identified by two independent personal identifiers of which at least one should be readable without technical equipment (e.g. barcode readers).

2.7 Transport of diagnostic material

Because diagnostic material should be considered as potentially infectious, transport outside the laboratory and the hospital is subject to national and international regulations. For road-bound transport in Europe, the European Agreement concerning the International Carriage of Dangerous Goods by Road (ADR) applies. In parallel, regulations issued by the International Air Transport Association (IATA) apply for the air-bound transport of diagnostic material.

2.7.1 *Category A Infectious Substances*

The list of *Category A Infectious Substances* includes infectious substances capable of causing permanent disability or life-threatening or fatal disease in otherwise healthy humans. Samples are assigned UN 2814 – *Infectious Substances* based on the known medical history and symptoms of the human source, endemic local conditions, and professional judgment concerning individual circumstances. If there is any doubt as to whether or not a pathogen may fall within this category, it should be transported according to the *Category A Infectious Substance* transport regulations.

For transportation of any *Category A Infectious Substance*, packaging that meets *UN* performance requirements for *Class 6.2* substances as shown by design type testing must be used. This is known as *UN* type-approved packaging and is certified and labeled accordingly. The packaging must include an inner packaging, an itemized list of contents, and an outer packaging. The inner packaging must comprise one or more leak-proof primary receptacle(s) and leak-proof secondary packaging. For liquid specimens, an absorbent material in sufficient quantity to absorb the entire contents must be placed between the primary receptacle and the secondary packaging; if multiple fragile primary receptacles are placed in a single secondary packaging, they must be either individually wrapped or separated to prevent contact between them. The primary receptacle(s) or the secondary packaging must be capable of withstanding without leakage an internal pressure producing a pressure differential of 95 kPa and temperatures in the range −40°C to +55°C. Furthermore, the inner package must be capable of successfully passing a drop test from a height of 1.2 m. There must be no leakage from the primary receptacle(s) and these must remain protected by the absorbent material if required.

Between the secondary packaging and the outer packaging, an itemized list of contents must be enclosed. Finally, a rigid outer packaging of adequate strength for its capacity, mass, and intended use must be added with the smallest external dimension being not less than 100 mm. The outer packaging may contain refrigerants such as dry ice. In that case, the primary receptacle(s) and the secondary packaging must maintain their integrity at the temperature of the refrigerant used.

2.7.2 Category B Infectious Substances

The list of *Category B Infectious Substances* includes infectious substances that do not meet the criteria for inclusion in *Category A Infectious Substances*. Samples are assigned *UN 3373 – Biological Substances, Category B*. In clinical practice, the majority of diagnostic material will be assigned category B.

For transportation of any *Category B Infectious Substance*, packing instructions are essentially identical to those for any *Category A Infectious Substance*, except that there is no need to use formally *UN* type-approved packaging. The outer packaging must be clearly labeled on the external surface of the outer packaging using a diamond-shaped label (Fig. 2.1). The proper shipping name *Biological Substance, Category B* in letters at least 6 mm high must be written on the outer package adjacent to the diamond-shaped label.

Fig. 2.1: The label for the outer packaging when shipping *Category B Infectious Substances*.

2.7.3 *Exempt patient specimens*

Patient specimens with minimal likelihood of pathogens present may be transported as *Exempt Patient Specimens*. To determine whether a patient specimen has a minimal likelihood that pathogens are present, an element of professional judgment is required. The judgment must be based on the known medical history, symptoms and individual circumstances of the source, human or animal, and endemic local conditions. Examples of specimens which may be transported under this paragraph include blood or urine required to monitor organ function such as heart, liver or kidney function; to determine drug or alcohol levels; to perform pregnancy tests; to investigate biopsies to detect cancer; or to study antibody titers in humans or animals in the absence of any concern for infection.

For transportation of any *Exempt Patient Specimen*, the packaging must include one or more leak-proof primary receptacle(s), a leak-proof secondary packaging, and an outer packaging of adequate strength for its capacity, mass and intended use and with at least one surface having minimum dimensions of 100 × 100 mm. For liquids, absorbent material in sufficient quantity to absorb the entire contents must be placed between the primary receptacle(s) and the secondary packaging so that any release or leak of a liquid substance will not reach the outer packaging and will not compromise the integrity of the cushioning material. If multiple fragile primary receptacles are placed in a single secondary package, they must be either individually wrapped or separated to prevent contact between them. The outer packaging must be clearly labeled *Exempt Patient Specimen*.

2.8 Stability of nucleic acids of selected pathogens during preanalytics

2.8.1 Human immunodeficiency virus type 1 (HIV-1) RNA

The level of HIV-1 RNA in plasma serves as an indicator of total viral load and thus viral replication rates. Quantitation of plasma HIV-1 RNA has been shown to be an important marker of disease progression, antiretroviral therapy efficacy, and risk of perinatal transmission. Standardized laboratory assays for HIV-1 RNA quantitation help to decrease both intra- and interlaboratory variability which is essential for

Tab. 2.3: Stability of HIV-1 RNA.

Specimen/sample	Temperature °C	Stability assured
EDTA whole blood	25	72 h
EDTA plasma	25	7 days
EDTA plasma	2–8	14 days
EDTA plasma	−20	5 years

optimal patient management. However, preanalytical conditions may have a significant impact on the results obtained.

For quantification of HIV-1 RNA, EDTA plasma clearly represents the material of choice. With citrate plasma, only 86% of HIV-1 RNA copies and with heparin plasma, only 69% HIV-1 RNA copies are yielded when compared with EDTA plasma. While the reduced yield when using citrate plasma can be explained by the dilution effect, the lower yield when using heparin plasma can be explained by the inhibitory effect of heparin. Serum is not feasible for HIV-1 RNA quantification because of clot formation; only approximately 25% of the viral load is yielded compared with quantification using EDTA plasma.

When collected in EDTA whole blood tubes, HIV-1 RNA can be stable for up to 72 h at room temperature; however, it is recommended to separate plasma through centrifugation within 12 h to prevent PCR inhibition due to hemolysis (Tab. 2.3). After centrifugation, EDTA plasma can be stored for up to 7 days at room temperature or up to 14 days at 4°C without significant decrease of the viral load measured. Storage at −20°C prevents the decline of the viral load measured for at least 5 years. Long-term storage should be at −70°C or lower. Use of tubes with gel separator, which ensures stable separation of EDTA plasma without opening, is strongly recommended.

2.8.2 Hepatitis B virus (HBV) DNA

Quantitation of HBV DNA in serum or EDTA plasma serves to determine eligibility for antiviral therapy and to monitor treatment response. Standardized commercial assays for quantitation of HBV DNA exist for more than a decade; however, only few studies have been performed to evaluate sample stability. Whereas some state that HBV DNA may be stable for at least 28 days up to 25°C, others recommend separation of serum or EDTA plasma within 4 h and storage at 4°C (Tab. 2.4). Use of tubes with gel separator, which ensures stable separation of serum or EDTA plasma without opening, is strongly recommended. Repeat freeze–thaw cycles should be avoided, although it has been reported that up to eight freeze–thaw cycles do not appear to have a significant effect on HBV DNA.

Tab. 2.4: Stability of HBV DNA and HCV RNA.

Specimen/sample	Temperature °C	Stability assured
Whole blood/EDTA whole blood	25	6 h[a]
Whole blood/EDTA whole blood	2–8	24 h
Serum/EDTA plasma	25	24 h
Serum/EDTA plasma	2–8	7 days
Serum/EDTA plasma	–20	5 years (HBV DNA), 1.2 years (HCV RNA in EDTA plasma)

[a] Insufficient data.

2.8.3 Hepatitis C virus (HCV) RNA

The presence of HCV infection is confirmed through detection of HCV RNA. Quantitation of HCV RNA in serum or EDTA plasma serves to determine eligibility for antiviral therapy and to monitor treatment response. EDTA plasma is the preferred specimen over heparinized plasma which may show inhibitory effects. HCV RNA is susceptible to degradation through RNase present in blood; therefore, whole blood specimens should be centrifuged as soon as possible, at least within 6 h (Tab. 2.4). Notably, the decay of HCV RNA in whole blood is more pronounced at 4°C than at room temperature; lysis of granulocytes may play a major role. Use of tubes with a gel separator, which ensures stable separation of serum or EDTA plasma without opening, is strongly recommended. Repeat freeze–thaw cycles should be avoided, although samples seem to be stable for up to three freeze–thaw cycles.

2.8.4 *Chlamydia trachomatis* and *Neisseria gonorrhoeae* DNAs

Both *C. trachomatis* and uncomplicated *Neisseria gonorrhoeae* infection are usually confined to the mucosa of the cervix, urethra, rectum, or throat. Both infections are often asymptomatic among females and, if untreated, may produce severe long-term effects.

For collection of swabs, commercially available collection kits employing specific stabilizing agents are used. If the swab is placed in the appropriate specimen transport tube at 25°C, it should be assayed within 60 days of collection (Tab. 2.5). At –20°C, swabs and urine specimens may be stored for up to 3 months after collection.

Tab. 2.5: Stability of *C. trachomatis* and *N. gonorrhoeae* DNAs.

Specimen	Temperature °C	Stability assured
Swab in specimen transport medium	25	60 days
Swab, urine	–20	3 months

2.8.5 Viral pathogens producing respiratory tract infections

Detection of viral pathogens producing respiratory tract infections is of major importance for patient management allowing for the appropriate use of drugs (avoidance
of the inappropriate use of antibiotics and initiating antiviral therapy if available),
and implementation of infection-control precautions. From a public health point of
view, accurate determination of the etiologic agent associated with an outbreak of
respiratory disease can be of value in managing the outbreak.

A variety of specimens including (induced) sputa, nasal swabs, nasopharyngeal
swabs, nasopharyngeal aspirates, throat swabs, and bronchoalveolar lavages are
suitable for testing on viral nucleic acids. After collection, swabs are placed in viral
transport media. Viral transport media usually consist of modified Hank's balanced
salt solution supplemented with bovine serum albumin, sucrose, glutamic acid, and
gelatin. The buffer maintains a pH value of 7.3 ± 0.2. Antibiotics and amphotericin
B are usually present to inhibit contaminants. Furthermore, viral transport media
usually contain cryoprotectants. Today, several collection devices including viral
transport media tubes are commercially available.

Specimens placed in viral transport media tubes are stable at room temperature
for 2 days; cooling at 2–8°C extends stability for up to 4 days (Tab. 2.6). For long-
term storage, freezing at –70°C is recommended. Recently, it has been suggested to
use dried swabs which turned out to ensure stable specimens for as long as 15 days
at room temperature. This simple and cost-effective sampling strategy seems to be
at least equally effective as conventional transport in commercially available viral
transport media.

2.8.6 Pathogens in stool specimens

No exhaustive studies exist that evaluate the stability of nucleic acids in stool specimens. It is advisable to collect the specimen in a sterile container and transport it
rapidly to the laboratory, if possible within 24 h at 4°C. Storage at room temperature
for more than 72 h may result in a significant viral decay and false-negative results.
Long-term storage at –20°C seems to be feasible until further testing. However, further

Tab. 2.6: Stability of respiratory virus nucleic acids.

Specimen	Temperature °C	Stability assured
Swab in viral transport medium	25	2 days
Swab in viral transport medium	2–8	4 days
Swab in viral transport medium	–20	2 months
Dried swab	25	15 days

studies concerning stability of nucleic acids in stool specimens are highly desirable to standardize collection and storage of stool specimens.

2.9 Take home messages

- Most errors occur in the preanalytical phase.
- Clear sampling instructions may help improving sample quality.
- In general, samples should be transported and stored at 4°C to avoid degradation of nucleic acids. Repeated freeze–thawing cycles should be avoided.
- Samples are usually stable for at least 24 h.
- To avoid cross contamination, PCR testing should be performed out of the unopened primary tube.

2.10 Further reading

[1] IATA. Guidance Document Infectious Substances. IATA; 2009 [cited 2009 12.10.2009]; Available from: http://www.iata.org/NR/rdonlyres/9C7E382B-2536-47CE-84B4-9A883ECFA040/0/ Guidance_Doc62DGR_50.pdf.
[2] UNECE. European Agreement concerning the International Carriage of Dangerous Goods by Road. United Nations Economic Commission for Europe 2007 [cited 2009 12.10.2009]; Available from: http://www.unece.org/trans/danger/publi/adr/adr2007/07ContentsE.html.

3 Quality assurance and quality control

Reinhard B. Raggam, John Saldanha and Harald H. Kessler

After meeting the requirements regarding preanalytics, specific considerations and actions concerning quality assurance and quality control issues must be taken. Technological improvements, from automated sample preparation to real time amplification technology, provide the possibility to develop and run assays for most of the pathogens of clinical interest. In contrast, quality assurance and quality control issues have often remained underdeveloped or are highly diverse. Furthermore, only a limited number of international standards and reference materials are available. This chapter focuses on the general requirements and specific components needed for validation and verification work in the routine molecular diagnostics laboratory for infectious diseases.

3.1 Accreditation issues

The framework of common standards including the international standards ISO 9001:2008 and ISO 15189 have been established allowing laboratories to plan and operate medical testing with an effective quality management system that has strong elements of quality assurance, quality control and quality improvement. When medical testing laboratories effectively implement this quality management system, they have continuous assurance that they are meeting their customers' needs and expectations for consistent, accurate, and timely test results.

Among several issues, these standards contain certain procedures for validation and verification work because the suitability of a laboratory technique does not necessarily prove that it will be performed correctly and provide valid results (Fig. 3.1). To prove this, quality management systems have been implemented in the majority of routine diagnostic laboratories. For laboratories in the European Union, the European Union's Directive on *In Vitro Diagnostic* (IVD) *Medical Devices* (98/79/EC) requires data demonstrating that an IVD achieves the stated performance and will continue to perform properly after it has been shipped, stored and put to use at its final destination. Additionally, common technical specifications enforced for tests or test systems are outlined in the Commission Decision of 7 May 2002 on common technical specifications for IVD medical devices. For laboratories in the US, the Food and Drug Administration (FDA) has established regulations based on current ISO standards. In addition to supervision, description and conformity of processes by the use of standard operating procedures, validation work focuses on the competence of the laboratory providing reliable test results and their correct interpretation.

```
┌─────────────────────────────────────────────────────────┐
│        Validation work (ISO 9001:2008; ISO 15189)         │
└─────────────────────────────────────────────────────────┘
                            │
┌─────────────────────────────────────────────────────────┐
│     Implementation of a new molecular test or test system │
└─────────────────────────────────────────────────────────┘
                            │
┌─────────────────────────────────────────────────────────┐
│                    Verification work                      │
└─────────────────────────────────────────────────────────┘
```

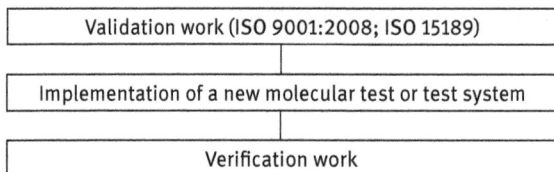

Fig. 3.1: Validation and verification.

As for all medical testing, laboratories performing molecular assays must adhere to established validation practices to ensure confidence and reliability in test results produced. It is worth noting, that the IVD Directive 98/79/EC includes not only the definition "test" but also the definition "test system" if more than a single component are required to generate a diagnostic result. For instance, molecular assays based on PCR usually consist of a combination of different reagents and instruments for nucleic acid extraction, amplification, and detection of amplification products. Only tests or test systems with proven suitability may be used in the routine molecular diagnostic laboratory, demanding verification work for each test or test system.

3.2 Standardization of diagnostic tests or test systems

Standardization of diagnostic tests or test systems for infectious disease markers is necessary to facilitate clinical guidelines used in the diagnosis, treatment, and monitoring of patients. The major goal of standardization is to ensure that the same numeric result for an analyte, e.g. nucleic acid, will be obtained for a sample irrespective of the assay method used to derive that result. One of the major issues with standardization of a diagnostic test is the availability of well-characterized, standardized reference materials. At present, very few reference materials are available for the numerous molecular diagnostic tests/test systems. The majority of tests or test systems use either an internally-developed reference reagent or a commercial material. Since the characterization of such materials is limited to a few tests or test systems, comparison of results from different tests/test systems is difficult. In addition, the concentration of the analyte may be expressed in different units, which also contributes to the difficulty in interpretation of results. For example, the concentration of nucleic acid in a sample is sometimes expressed in copies/ml, genome equivalents/ml, or PCR detectable units/ml.

Molecular diagnostics usually utilize clinical samples which are complex mixtures of biological materials where the analyte is poorly defined. Typically, independent measurement procedures or reference laboratories are not available and the measurement cannot be traced to an SI unit. Since the tests are based on a biological function (e.g. nucleic acid amplification), the unit defining such materials, the International

Unit (IU), is arbitrary but provides a common biological unit of measurement. The establishment of WHO International Standards, primary reference materials against which reference reagents (secondary, tertiary, run controls, etc.) are calibrated, has provided global reference materials which enable direct comparisons between products and measurements across different methodologies and tests or test systems.

In order to establish an International Standard, an international collaborative study is done, involving as many different tests or test systems as possible so that the final assignment of the material is test- or test system-independent. The composition of the active material in an International Standard resembles, as closely as possible, that of samples to be analyzed. The candidate materials typically include clinical specimens but may include, for example, cell-culture derived viruses for NAT. The final titer assignment of an International Standard, in IUs, is based on the results of the collaborative study and the agreement of the WHO Expert Committee on Biological Standardization (ECBS). Since the titer of an International Standard is assigned, there is no uncertainty of measurement associated with any International Standard or subsequent replacement International Standards. Generally, for viral NAT standards, the most common genotype is used. However, this may cause issues, especially with NAT tests that have not been optimized for all the known genotypes of a virus. In such cases, genotype panels are available for the common blood borne viruses and can be used to determine the genotype inclusivity of a test or test system.

The first WHO International Standards for NAT were established in the mid-1990s for qualitative NAT used by plasma manufacturer standardization of results across test systems and enabled regulators to set corresponding NAT guidelines. Quantitative test systems for determination of viral load were first established for HIV-1 and subsequently for other viruses such as HCV and HBV. In the last decade, the accuracy and precision of these test systems has continued to improve. This is of major importance since even very small variations in the value assignment between an International Standard and its replacement could be problematic, especially for viral load testing where treatment guidelines are based on accurate quantitation. In addition, the International Standards represent the highest order standard available and are used to calibrate a range of secondary reference materials. Therefore, in order to minimize the regular replacement of International Standards, secondary standards, working reagents and calibrators, directly traceable to the WHO International Standards, should be used for all routine procedures. International Standards and genotype panels are available from several organizations including the National Institute for Biological Standards and Control, UK (http://www.nibsc.ac.uk), Paul Ehrlich Institute (PEI), Germany (http://www.pei.de), and the Food and Drug Administration (FDA), USA (http://www.fda.gov).

A recent issue that has arisen is the commutability of the International Standards. Commutability is defined as the ability of a reference or control material to have inter-assay properties comparable to properties demonstrated by clinical samples when measured by more than one analytical method. However, many reference materials currently used are not prepared from a single clinical sample but may be derived from a pool of samples, material purified from its native matrix, or cell culture material and are not commutable with clinical specimens. When such materials are used as calibrators, different diagnostic tests or test systems do not give comparable results with patient samples. The laboratory must therefore include validation of commutability of reference materials in order to demonstrate suitability for intended use.

3.3 Validation and verification work

Validation and verification work is defined as confirmation through the provision of objective evidence that requirements for a specific intended use or application have been fulfilled (ISO 9001:2008). Components of validation work must be applied to ensure that a procedure, process, system, equipment, or method used works as expected and achieves the intended result constantly (WHO-BS/95.1793). In contrast, components of verification work are assigned to determine or confirm the performance characteristics of a molecular test or test system before it is used for patient testing.

3.4 Components of validation work

According to ISO 9001:2008 and ISO 15189, components of validation work for a molecular test or test system include the implementation of an appropriate quality control system. This consists of both internal and external quality controls and the use of international standards and reference materials if available. Further requirements include validation of the employee competency, calibration procedures of the instruments used, and monitoring of the test results achieved in correlation with clinical findings regarding the diagnostic sensitivity and the diagnostic specificity (Tab. 3.1).

3.4.1 Internal and external quality controls

Because amplification may fail due to interference from inhibitors, an internal control (IC) must be incorporated in every molecular test or test system to exclude false-negative results. ICs are needed to exclude false-negative results due to interference from inhibitors. To ensure reliable test results, the IC must be added to the sample before the start of the nucleic acid extraction procedure. Either a homologous or a heterologous IC can be used. The homologous IC is a DNA sequence (for DNA amplification targets) or an *in vitro* transcript (for RNA targets) consisting of primer binding regions identical to those of the target sequence. It consists of a randomized internal sequence with a length and base composition similar to those of the target

Tab. 3.1: Components required for validation work of molecular tests or test systems.

Component required	Particular requirements
Quality control: internal controls, external run controls, international standards and reference materials	Quality control records including performance data, instrument printouts, and corrective actions
Proficiency testing participation for comparison of inter-laboratory test results	Proficiency testing results including corrective actions
Validation of employee competency	Key and operating staff qualifications, credentials verifying field expertise
Instrument maintenance and calibration	Instrument records including printouts /digital archiving of maintenance and calibration protocols
Correlation with clinical findings	Diagnostic sensitivity and specificity of the test or test system

sequence and a unique probe-binding region that differentiates the IC amplification product from the target amplification product (Fig. 3.2). Either a single IC or multiple ICs for a set of molecular assays can be generated. In contrast to the homologous IC, the heterologous IC represents a second amplification system within the same reaction vessel (Fig. 3.3). The control must have the same or similar extraction and amplification efficiencies as the target. Plasmids or housekeeping genes can be used as heterologous ICs. Notably, any IC (homologous or heterologous) should be added at a concentration close to the limit of detection/quantitation to ensure successful removal of inhibitors on the one hand and to prevent competition with the target template for reagents on the other.

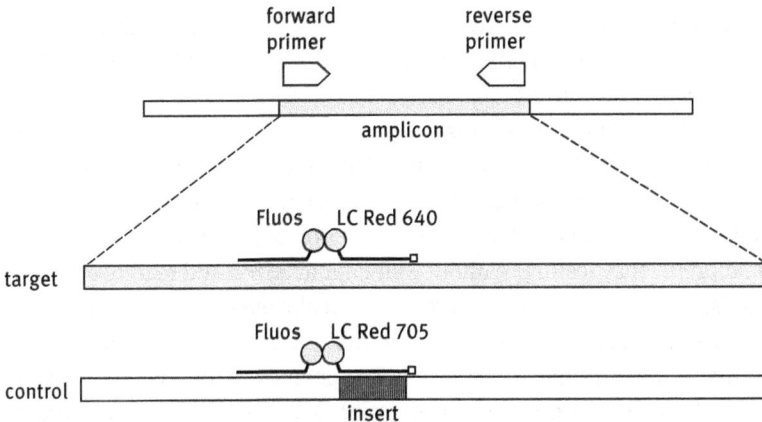

Fig. 3.2: Homologous internal control showing an *in vitro* DNA transcript. The primer annealing regions are identical to those of target sequence. One probe binding region is identical to that of the target sequence; the second probe binding region is different to that of the target sequence (probe labeled with a different fluorescence dye).

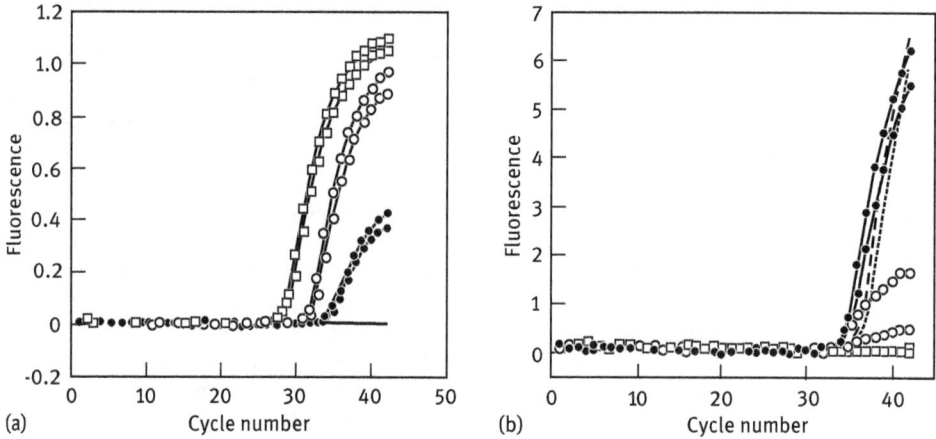

Fig. 3.3: Real-time PCR including a heterologous internal control (human beta microglobulin serving as "housekeeping gene") by using a second amplification system in parallel in the same reaction vessel but with different fluorescence dyes. (a) Target amplification curves (run in duplicate). (b) Internal control amplification curves. Note that the amplification of the internal control is competitively inhibited through that target with the highest concentration (squared curves).

To monitor the correctness of a test result obtained with the new molecular test or test system continuously after implementation in the routine diagnostic laboratory, the introduction of an external run control (ERC) which is independent from the external positive control(s) included by the manufacturer is recommended. Because the ERC must monitor the whole molecular assay including sample preparation, the ERC is added directly into the sample prior starting with nucleic acid extraction. An ERC may be implemented either in each test run or within defined intervals, e.g. when introducing a new test lot. Comparison of the results obtained by the ERC with those obtained by the external positive control(s) included by the manufacturer enables identification of relevant aberrations at an early stage (Fig. 3.4).

An integral part of quality assurance is the use of well-characterized and readily available reference materials, to maintain clinical laboratory quality assurance for molecular tests or test systems testing. Reference materials can be used for quality control, verification work for tests or test systems, detection of errors, monitoring of test performance, and proficiency testing. Without well-characterized and readily available reference materials, it is difficult to cross-reference molecular assays, IVD/CE-labeled or FDA-approved or -cleared molecular tests as well as laboratory-developed ones.

Recently, a hierarchy of reference materials has been described which is based on the degree of characterization of each material. The top category consists of *ISO Reference Materials*. These materials are sufficiently homogenous and stable with respect to one or more specified properties. They have been established for their intended use in a measurement process. *ISO Reference Materials* have been provided by organizations including the WHO as independent organization and

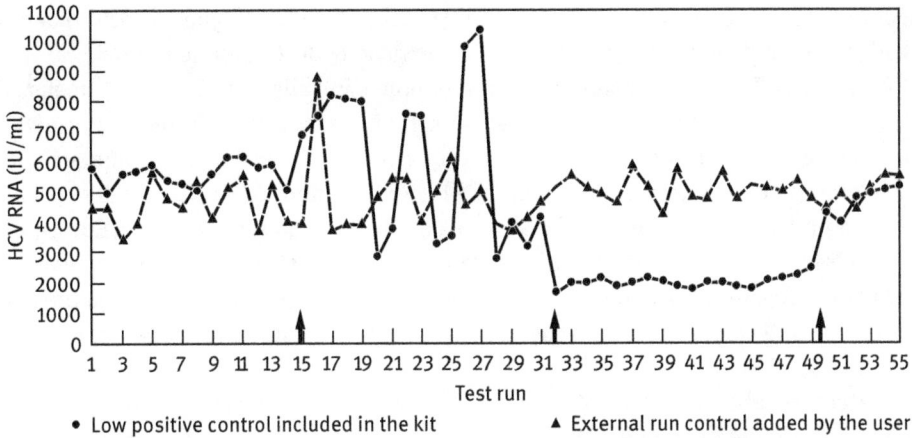

Fig. 3.4: Performance of an external run control implemented in a molecular assay for quantification of hepatitis C RNA (arrows indicate introduction of a new test lot).

several commercial distributors including the Institute of Reference Materials and Measurements. The second category consists of *ISO Certified Reference Materials*. These materials are characterized by a metrological valid procedure for one or more specified properties, accompanied by a certificate that provides the value of the specified property, its associated uncertainty, and a statement of metrological traceability. *ISO Certified Reference Materials* are provided e.g. by the National Institute of Standards and Technology. They include commercially available reference materials, which either are CE-labeled (Conformité Européenne, a mandatory conformity mark for products placed within the European Union market) or approved or cleared for *in vitro* diagnostic use by the FDA.

In general, the number of reference materials developed by governmental organizations and manufacturers, which can be used for nucleic acid-based pathogen detection, is too small. This lack of widely available reference materials enforces the laboratories to use other types of materials as controls such as the remainder of a patient specimen that was left following clinical routine testing. In addition, because reference materials are generally available in limited quantities, laboratories will need more widely available materials in the future that can be used on a daily basis for a variety of purposes, including quality controls.

3.4.2 Proficiency testing

Proficiency testing is a continuous process for checking actual laboratory testing performance, usually by means of inter-laboratory test results comparisons. Each medical testing laboratory must enroll in an approved proficiency program or programs, for each of the specialties and subspecialties for which it seeks accreditation.

For example, to assess a laboratory's ability to use molecular diagnostic technologies within the clinical setting, taking part on proficiency testing panels providing the laboratory with a series of samples that resemble clinically significant specimens is mandatory. In Europe, several programs for proficiency testing in molecular diagnostics for infectious diseases have been established including several commercially available programs provided by e.g. QCMD, NEQAS, and INSTAND and by *QUALqual*, a project proposed under the auspices of the EC4 and funded by the European Commission. Results obtained from these programs reveal that a higher percentage of correct results is usually observed with IVD/CE labeled and/or FDA-approved or-cleared molecular tests or test systems than with laboratory-developed molecular assays.

Medical laboratories must be able to provide the proficiency testing results together with the individual reports to the national body of laboratory accreditation. Data must include test runs with results, reports, report lists, and signatures. A singular poor score in a certain proficiency-testing program does not mean that the test or test system must be abandoned immediately. However, immediate action must be taken to overcome the existing deviation(s). The unsuccessful correction of the existing deviation(s) can lead to suspension of the test or test system affected by the national body of laboratory accreditation.

3.4.3 Validation of employee competency

Both key and operating laboratory staff must remain competent in performing molecular tests and test systems and in reporting valid results. A large number of methods are available for the acquisition of the specific knowledge. There are numerous providers offering training on molecular diagnostics. The laboratory staff must participate and complete the training successfully, yielding credentials that verify field expertise on a constant level. These procedures eventually help to ensure the consistency of the results produced and reported by the molecular diagnostics laboratory.

3.4.4 Instrument maintenance and calibration

Instrument calibration, instrument maintenance and function checks contribute to the on-site assessment of a medical testing laboratory. To comply with relevant standards, a documented and recorded program of maintenance and calibration following the manufacturer's recommendations is required. Additionally, the manufacturer's instructions, the operator's manuals, and updates of documentation must be used to meet defined criteria (e.g. frequency and modification of procedures) regarding maintenance and calibration.

3.4.5 Correlation with clinical findings

The implications of a false-negative or a false-positive test result and the impact of a laboratory result on the diagnosis and management of the disease require consideration regarding the diagnostic sensitivity and the diagnostic specificity of a test or test system. The diagnostic sensitivity of a test or test system is determined by the proportion of patients with well-defined clinical disorders whose test values are positive or exceed a defined limit of decision (i.e. a positive test result and identification of the patients who have the disease). It must be considered carefully that the clinical disorder must be defined by criteria independent of the test or test system used. The term *diagnostic sensitivity*, which is mainly used in the EU, is equivalent to *clinical sensitivity* mainly used in the US. The diagnostic specificity of a test or test system is defined by the ability of a measurement procedure to measure solely the analyte, thus avoiding false positive test results (i.e. a negative test result and identification of the patients where the disease is absent). The term *diagnostic specificity* is used uniformly in the EU and in the US.

A greater variability regarding diagnostic sensitivity and diagnostic specificity is usually observed with laboratory-developed molecular assays than with IVD/CE labeled molecular assays. Furthermore, it is difficult to cross-reference molecular assays, IVD/CE labeled as well as laboratory-developed ones, without well-characterized and readily available reference materials (see Section 3.3.1).

3.5 Components of verification work

Following successful implementation of the components of validation work, verification work for molecular tests or test systems must be performed. Because clear and comprehensive guidelines on verification work of molecular tests or test systems are lacking, a great diversity of approaches exists. This diversity is not only seen within the EU in general, but also often even within certain countries. Therefore, a common national or preferably an international regulatory framework on the verification work of molecular tests or test systems is still urgently needed. Suggestions described here may be a helpful approach towards a clear and meaningful verification work when implementing a new molecular test or test system in the routine diagnostic laboratory (Fig. 3.5).

Components of verification work of a molecular test or test system should depend on the type of assay used in the laboratory. For an IVD/CE labeled and/or FDA-approved or -cleared test or test system, the manufacturer is responsible for the IVD achieving the performance as stated. However, it is advisable to determine or confirm the performance characteristics prior to the use for patient testing. In contrast, for a

Validation (ISO 9001:2008; ISO 15189)

Implementation of a new molecular test or test system

Verification work

IVD/CE; FDA	Laboratory-developed

Accuracy Imprecision Analytical measuring range Recovery (interference)	Accuracy Imprecision Analytical measuring range Recovery (interference) Reproducibility Specificity LOD/LOQ

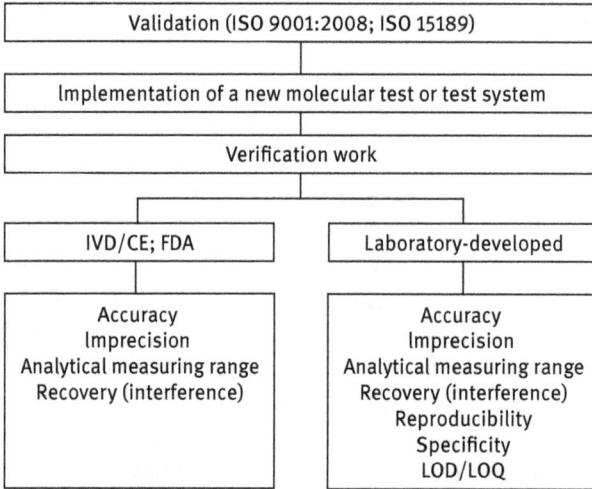

Fig. 3.5: Workflow for implementing a new molecular test or test system in the routine diagnostic laboratory.

laboratory-developed test or test system, a research-use-only (RUO) test or test system, or the combination of different IVD/CE labeled or that of different FDA-approved or-cleared tests without recommendation of the manufacturer, or any change of a compulsory test procedure, the laboratory is responsible for both the suitability and the correct performance of the test or test system demanding for an extended verification work before it is used for patient testing.

3.5.1 Components of verification work for IVD/CE labeled and/or FDA-approved or -cleared tests or test systems

This includes testing of accuracy, within-run and between-run imprecision, and in the case of a quantitative assay determination of the analytical measuring range (linearity). If another sample matrix is intended to be used than that meeting the manufacturer's specifications, recovery testing to rule out possible interfering substances should be performed additionally (Tab. 3.2).

3.5.1.1 Accuracy
The term *accuracy*, in its metrological sense, refers to the closeness of the agreement between the result of a (single) measurement and a true value of an analyte, and comprises both random and systematic effects (Fig. 3.6). Accuracy can be estimated by analyses of accepted reference materials or a comparison of results with those obtained by a reference method. These are the only accepted approaches to accuracy. When neither is available, other evidence is required to record the ability of

Tab. 3.2: Components required for verification work for IVD/CE-labeled and/or FDA-approved/-cleared tests or test systems.

Component required	Assigned value of the accepted reference material
Accuracy	Positive[a]
	Low positive[b]
	Negative
Within-run and between-run imprecision	Positive[a]
	Low positive[b]
Analytical measuring range	Positive[a]
Recovery (interference testing)	Positive[a]

[a] More than 1 $\log_{10}$ above the LOD and within the analytical measuring range.
[b] Up to 1 $\log_{10}$ above the LOD.

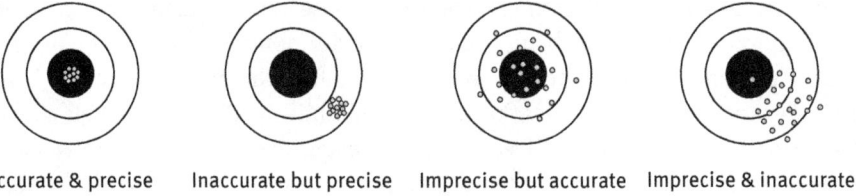

Accurate & precise Inaccurate but precise Imprecise but accurate Imprecise & inaccurate

Fig. 3.6: Accuracy and imprecision.

the method to measure the analyte. For determination of the accuracy, positive, low positive and negative samples are used.

3.5.1.2 Imprecision

Imprecision is defined as the level of deviation of the individual test results within a single run (within-run imprecision) and from one run to another (between-run imprecision). Imprecision, which is the dispersion of results of measurements obtained under specified conditions, is usually characterized in terms of the standard deviation of the measurements and relative standard variation, the coefficient of variation. For determination of imprecision (within-run and between-run) of a molecular test or test system, positive and low positive samples are used. In order to optimize the verification workflow, it may be useful to take the first result of within-run imprecision testing as the first result of between-run imprecision testing thus allowing a reduction in the number of experiments.

3.5.1.3 Analytical measuring range (linearity)

In the case of a quantitative molecular test or test system, the analytical measuring range (linearity) must be verified. The analytical measuring range is defined by a set of values of an analyte for which the error of a measuring instrument is intended to lie within specified limits. For determination of the analytical measuring range, a serial dilution of a positive sample is analyzed in duplicate (Fig. 3.7). Within the analytical measuring range, results are valid without further dilution.

3.5.1.4 Recovery (interference)

If a sample matrix is intended to be introduced for which the manufacturer has not verified the molecular test or test system, recovery testing should be performed additionally. The following experiments for verifying consistent recovery of analytes are accepted to rule out possible interfering substances: in the first experiment, interferences can be assessed by running a 1:2 and a 1:4 mixture of the newly introduced sample matrix with the diluent specified for use with the method. The laboratory should assess if there is comparability between the results for the two diluted samples and the neat result similar to that observed for the method using the verified sample matrix. If the results demonstrate dilution ratios and coefficients of variation are comparable to the verified sample matrix, the probability of an interfering substance affecting the neat result appears unlikely. If a reproducibly low or absent concentration is obtained with the first experiment, the newly introduced sample matrix should be used as the diluent for a verified sample matrix (routine patient sample) with a high measurable concentration in another set of 1:2 and 1:4 proportions in a second experiment. The effect of the newly introduced sample matrix on the neat result for the verified sample matrix (patient sample) is verified. If the results of the second experiment demonstrate dilution ratios and variability comparable to routine dilution, the neat newly introduced sample matrix is unlikely to exert an interfering effect. In contrast, if no acceptable results are obtained by the second

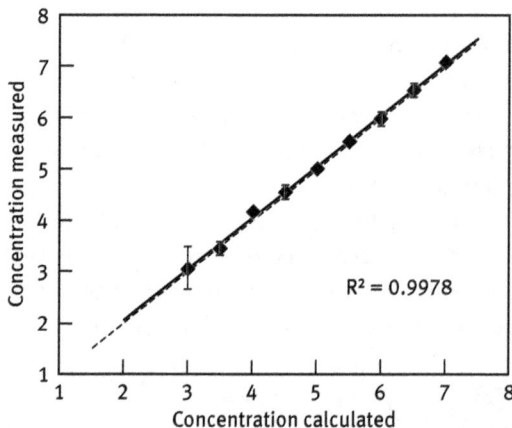

Fig. 3.7: Analytical measuring range obtained by dilution (0.5 $\log_{10}$ steps) of a positive sample.

experiment, this molecular test or test system appears not to be suitable for the newly introduced sample matrix.

Notably, the IC (see Section 3.3.1) checks for a possible matrix-induced effect and helps to ensure the reliability of a molecular test or test system.

3.5.2 Components of verification work for laboratory-developed tests or test systems

To determine or confirm performance characteristics of a laboratory-developed test or test system, a RUO test or test system, or the combination of different IVD/CE labeled tests or that of different FDA-approved or -cleared tests without recommendation of the manufacturer, additional components of verification work must be performed prior to the use for patient testing. Additional components include testing of reproducibility, specificity and, for a quantitative molecular test or test system, the determination of the limit of detection (LOD) and the lower limit of quantitation (LLQ).

3.5.2.1 Reproducibility
The reproducibility (of results of measurements) is defined as closeness of agreement between the results of measurements of the same analyte carried out under changed conditions of the measurement. If conditions such as the operator working with the test or test system, the test or test system itself, test or test system storage conditions and/or the location where the test or test system is performed within the laboratory change, verification of reproducibility is required. The closeness of the agreement of test results obtained is reported as standard deviation.

3.5.2.2 Specificity
Specificity testing checks for the ability of a measurement procedure to measure solely the analyte. Specificity ensures avoidance of false-positive results. It has no numerical value in this context. When establishing a laboratory-developed molecular test or test system, primer and probe sequences must be checked carefully by the use of a genome sequence databank. It is advisable to verify the amplification product by means of sequencing and to use a primer pair that has already been published in a highly recognized journal. The latter helps to avoid testing of a more or less extended specificity panel. Determination of specificity can be achieved by testing accepted reference materials or specificity panels tailored to its intended use.

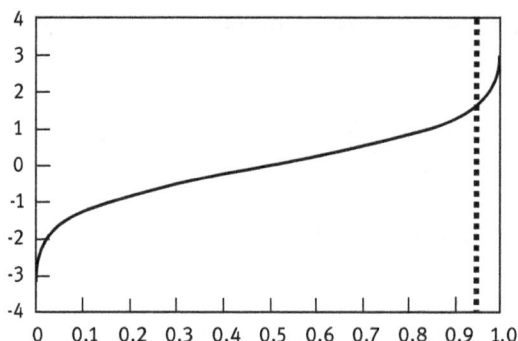

Fig. 3.8: Determination of the limit of detection through Probit analysis. The dotted line indicates the lowest concentration of an analyte that can be distinguished from the absence of that analyte within a stated confidence limit (95% recovery rate).

3.5.2.3 Determination of the limit of detection (LOD) and the lower limit of quantitation (LLQ)

The LOD is defined as the lowest quantity of an analyte that can be distinguished from the absence of that analyte (a blank value) within a stated confidence limit, usually the 95% recovery rate for a molecular test or test system, determined through Probit analysis (Fig. 3.8).

The LLQ is the lowest concentration at which the analyte can not only be reliably detected but also at which some predefined goals for bias and imprecision are met. The LLQ may be equivalent to the LOD or it could be at a (much) higher concentration. This depends on the specifications for bias and imprecision defined for the assay. Determination of both the LOD and the LLQ strongly depends on the availability of accepted reference materials (see Section 3.3.1). If there is no accepted reference material available, it is impossible to determine the LOD and the LLQ reliably.

3.6 Take home messages

–   For validation work, harmonization among the framework of common standards (ISO 9001:2008 and ISO 15189) exists. All reference materials, calibrators, etc., used for test or test system validation and routine testing should be traceable to a higher order standard such as a WHO International Standard if such a standard has been established.
–   Test or test system validation should include validation of any reference materials used to ensure commutability of these materials with the routine samples tested.
–   No harmonization exists currently for verification work; however, verification is mandatory if any new molecular test or test system for patient testing is introduced.

- Verification work helps to ensure reliable test results and contributes to a better comparability of molecular tests and test systems.
- Laboratory-developed tests or test systems, RUO tests or test systems, the combination of different IVD/CE-labeled tests and that of different FDA-approved or-cleared tests without recommendation of the manufacturer require additional components of verification work prior to the use for patient testing.

3.7 Further reading

[1] Commission Decision of 7 May 2002 on common technical specifications for in vitro-diagnostic medical devices (2002) Official J. Eur. Commun. L131:17–30.
[2] Directive 98/79/EC of the European Parliament and of the Council of 27 October 1998 on in vitro diagnostic medical devices (1998) Official J. Eur. Commun. L331:1–37.
[3] Greenberg, N. (2014) Update on current concepts and meanings in laboratory medicine – Standardization, traceability and harmonization. Clin. Chim. Acta. In press.
[4] Linnet, K., Boyd, J.C. (2006) Selection and analytical evaluation of methods – with statistical techniques. In: Burtis, C.A., Ashwood, E.R., Bruns, D.E., Tietz Textbook of Clinical Chemistry and Molecular Diagnostics, 4th edn., St. Louis: Saunders, 353–407.
[5] Madej, R.M., Davis, J., Holden, M.J., Kwang, S., Labourier, E., Schneider, G.J. (2010) International standards and reference materials for quantitative molecular infectious disease testing. J. Mol. Diagn. 12:133–143.
[6] Panteghini, M. (2009) Traceability as a unique tool to improve standardization in laboratory medicine. Clin. Biochem. 42:236–240.
[7] Payne DA, Mamotte CDS, Gancberg D, Pazzagli M, van Schaik RHN, Schimmel H, Rousseau F, on behalf of the IFCC Committee for Molecular Diagnostics (2010) Nucleic acid reference materials (NARMs): definitions and issues. Clin. Chem. Lab. Med. 48:1531–1535.

4 Extraction of nucleic acids

Harald H. Kessler

Extraction of nucleic acids is obligatory prior to amplification. The choice of an optimal extraction procedure is essential for the successful amplification and detection of nucleic acids. Failure to extract nucleic acids adequately may result in the inappropriate categorization of a specimen as falsely negative or falsely positive. To obtain correct results, the nucleic acid extraction protocol must warrant the efficient lysis of the nucleic acids-containing specimen and the removal of substances which might inhibit the subsequent steps of reverse transcription and/or amplification while protecting target DNA or RNA from degradation. Furthermore, cross-contamination caused by microorganisms must be avoided and potential hazards caused by toxic reagents should be kept to a minimum.

4.1 Manual nucleic acid extraction protocols

Classic laboratory developed manual nucleic acid extraction protocols were usually time-consuming, labor intensive, and susceptible to contamination. It is well known that the probability of false-positive results because of contamination increases in relation to the number of manipulations involved in sample processing. Furthermore, those protocols involved hazardous reagents such as phenol and chloroform. The basic principle included phenol–chloroform extraction, either with or without proteinase K digestion, and alcohol precipitation. If the ethanol for the precipitation of nucleic acids was not properly removed, excess ethanol residues were responsible for inhibition of subsequent amplification. RNA extraction protocols were even more laborious than those for DNA were. Although only specially trained staff were employed, a technician-dependent variability in the efficacy of extraction could be observed and contamination because of numerous manipulations involved was a major concern. Those protocols were thus largely replaced by commercially available kits.

Several companies brought prepackaged manual nucleic acid extraction kits to the market. They usually employ silica gel membranes, glass fiber surfaces, magnetic beads, or chaotropic salts with or without silica gel columns. These kits are more rapid to perform and include fewer manipulation steps. They have largely replaced the classic laboratory developed manual nucleic acid extraction protocols; however, they are still far from being useful in the routine diagnostic laboratory.

4.2 Automated nucleic acid extraction platforms

Widespread use of molecular assays in the routine diagnostic laboratory strongly depends upon automation. Recently, several commercially available ready-to-use extraction kits on automated platforms were brought to the market. They provide significantly decreased hands-on time, avoid potential hazards caused by toxic reagents, and show significantly increased convenience and ease of use. Ideally, the automated platform allows hands-off operation from the moment of loading the patient specimen up to the completed nucleic acid extract permitting the operator to do other duties. A major concern in the implementation of automated nucleic acid extraction platforms was the potential for cross-contamination because of aerosols, pipette leaks, faulty robotics, or robotic error. However, latest generation platforms keep sample manipulation to a minimum and many of them provide closed conditions; evidence of cross-contamination has not been observed thus far. In the routine diagnostic laboratory, commercially available automated platforms for nucleic acid extraction provide increased reliability and repetitiveness in comparison with manual nucleic acid extraction protocols.

4.2.1 Technology principle

Today, the majority of nucleic acid extraction protocols on automated nucleic acid extraction platforms are based on the magnetic glass (also called silica) particle technology. This technology consists of four major steps: lysis, capture, purification, and elution (Fig. 4.1). In the first step, a certain amount of lysis buffer is added to the sample. The specially composed lysis buffer is designed to break up the pathogen, followed by the release but simultaneous stabilization of total nucleic acids (at high salt concentration and neutral pH value). Furthermore, it degrades inhibitory proteins and RNases through protease digestion and inactivation of nucleases through chaotropic salt. In the following step, released total nucleic acids are captured through binding to the silica surface of added magnetic glass particles in the presence

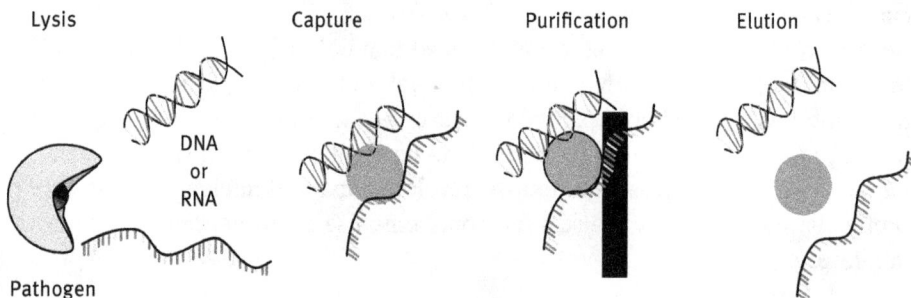

Fig. 4.1: The four major steps of the magnetic glass (also called silica) particle technology.

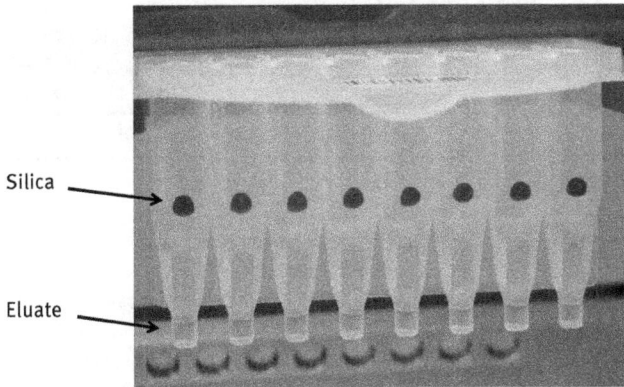

Fig. 4.2: Situation after the final step of nucleic acid extraction: the total nucleic acids have been recovered in a small amount of elution buffer (also called eluate).

of chaotropic salt (at high salt concentration and neutral pH value). Today, magnetic glass particles with a significantly increased silica surface have been brought to the market. After nucleic acids capture, purification is performed through repeat washing steps. The wash buffer employed removes unbound substances and impurities such as denatured proteins and cellular debris which may act as potential PCR inhibitors (at low salt concentration and low pH value). Finally, the purified total nucleic acids are released at elevated temperature (low salt concentration and high pH value) and recovered in a small amount of elution buffer (Fig. 4.2). This buffer keeps the nucleic acids in proper condition to secure them from degradation. The smaller the elution buffer volume has chosen, the higher the level of detection of nucleic acids.

4.2.2 Desirable features of automated platforms

The introduction of an automated platform in the routine diagnostic laboratory requires a closer consideration of general and specific features (Tab. 4.1). Besides the potential to extract nucleic acids within 1 to 2 hours, flexibility is a major issue. Automated platforms should be able to extract nucleic acids from a wide range of different types of specimens. They should allow adding variable input volumes. For example, for specimen types such as throat washings, bronchoalveolar lavages, or urine, the input volume may be increased compared to blood samples. Additionally, automated platforms should be able to generate variable elution volumes. Cerebrospinal fluid samples or specimens obtained from infants may only be available with low volumes demanding for a decreased elution volume. Notably, platforms ensuring the identical volumes for input, loading, and processing (without any dead volume) are preferable. Both extraction of DNA and RNA in a single run should be possible. Furthermore, the addition of internal controls prior to the start of the run should be feasible.

Tab. 4.1: Features to be considered when selecting an automated platform for nucleic acids extraction.

General features to be considered	Specific features to be considered
Dimension of instrument	Addition of internal controls prior to the start of the run
Low-throughput, medium-throughput, or high-throughput platform	Ready-to-use reagents in sealed containers/bottles
Flexibility (extraction of DNAs and RNAs in a single run, number of specimens, type of specimens, range of sample input volume, range of elution volume)	Disposables for each individual sample
Maintenance procedures	Closed/screw-cap reaction vessels
Safety measures (load check, label code reading)	Possibility of automated pipetting of PCR mixes into different types of PCR vessel but at least to a transparent 96-well plate
Hands-on time/turnaround time	Possibility of automated addition of eluates
Cost of instrument/disposables/reagents	

4.3 Preparation of qPCR mixes and addition of eluates (qPCR assay setup)

Due to the growing number of samples to be analyzed with different assays in the molecular diagnostic laboratory for infectious diseases, automated preparation of qPCR mixes and addition of eluates (qPCR assay setup) are highly desirable. However, for many applications, automated qPCR assay setup is impossible due to the lack of a suitable platform. The ideal platform must be able to extract nucleic acids and to prepare different qPCR mixes for detection of different pathogens (according to the specific order list) quickly in parallel. It must be adaptable for different molecular assays from different manufacturers. Similar to a clinical chemistry analyzer, the platform should be able to produce printable working lists. After completion of extraction, the ideal platform must be able to add eluates immediately to the qPCR mixes. It should provide a choice of output adaptors to enable use of different qPCR cyclers but at least be able to pipette into a transparent 96-well plate.

4.4 Currently frequently used commercially available platforms

Currently frequently used commercially available automated platforms for nucleic acid extraction and qPCR assay setup are shown in Fig. 4.3. Alternative platforms without the possibility of automated PCR assay setup are shown in Fig. 4.4. If one of the latter nucleic acid extraction platforms is used it may be useful to establish

Fig. 4.3: Currently frequently used commercially available platforms for nucleic acid extraction and qPCR assay setup. Upper row, from left to right: m2000sp (Abbott), QiaSymphony (Qiagen), and COBAS AmpliPrep (Roche). Lower row, from left to right: COBAS x 480 (Roche), VERSANT kPCR SP (Siemens), and SENTOSA SX101 (Vela).

Fig. 4.4: Currently frequently used commercially available platforms for nucleic acid extraction without the possibility of automated qPCR assay setup. From left to right: NucliSens easyMAG (bioMerieux), MagNA Pure LC 2.0 (Roche), and MagNA Pure Compact (Roche).

an additional platform for qPCR assay setup, e.g. the easySTREAM platform (bio-Merieux). Basic features of currently frequently used commercially available platforms for nucleic acid extraction are listed in Tab. 4.2. When introducing such a platform, it must always be considered whether the platform allows adaptation of protocols to the specific needs of the user ("open-mode") or it only allows nucleic acid extraction (and qPCR assay setup) as a front end of one or more molecular assays provided by the manufacturer ("closed-mode"). For platforms which are designed to work in an "open-mode", several manufacturers provide a "generic extraction protocol" allowing the possibility of extracting DNAs and RNAs from a broad range of different specimens such as serum, plasma, EDTA whole blood, cerebrospinal fluid, urine, sputum, and swabs in a single extraction run.

Tab. 4.2: Basic features of currently frequently used commercially available platforms for nucleic acid extraction and PCR assay setup (in alphabetical order). Note that turnaround times vary significantly depending on the number of different qPCR setups to be prepared.

Manufacturer	Platform	Sample input volume (µl)[a]	Elution volume (µl)[a]	Maximum run size
Abbott	m2000sp	400–1200	70–200	96
Qiagen	QiaSymphony	280–1100	60, 85, or 110	96
Roche	COBAS AmpliPrep[b]	250–1100	50 or 65	72
Roche	COBAS x480[b]	400	50	96
Siemens	VERSANT kPCR SP	225–625	50 or 100	96
Vela	SENTOSA SX101[b]	50–700	50–100	96

[a] Depending on protocol used.
[b] Only "closed-mode".

4.5 Take home messages

– Today, commercially available ready-to-use extraction kits on automated platforms are predominantly used in the routine diagnostic laboratory.
– Automated nucleic acid extraction platforms are attractive to users for a variety of reasons including increased reliability and decreased hands-on time.
– Automated platforms provide uniform and reproducible recovery of nucleic acids but it is of major importance to generate comparative data between different techniques.
– It must be considered that an extraction protocol which is used for one pathogen in a particular specimen type may not be useful for another pathogen in another specimen.
– Automated platforms for nucleic acid extraction and qPCR assay setup are highly desirable.

4.6 Further reading

[1] Berensmeier, S. (2006) Magnetic particles for the separation and purification of nucleic acids. Appl. Microbiol. Biotechnol. 73:495–504.
[2] Jungkind, D., Kessler, H.H. (2002) Molecular methods for the diagnosis of infectious diseases. In: Truant, A.L., ed. Manual of Commercial Methods in Clinical Microbiology, Herndon, VA: ASM Press.
[3] Verheyen, J., Kaiser, R., Bozic, M., Timmen-Wego, M., Maier, B.K., Kessler, H.H. (2012) Extraction of viral nucleic acids: comparison of five automated nucleic acid extraction platforms. J. Clin. Virol. 54:255–259.

5 Amplification and detection methods

Stephen A. Bustin and Harald H. Kessler

The last century has witnessed extraordinary progress in the battle against deaths caused by infectious diseases. The implementation of improvements in hygiene, sanitation and diet, together with universal vaccine programs to prevent childhood illnesses and the success of contemporary antibiotic and antiviral drugs has had dramatic effects. World-wide infant mortality has been reduced by half and life expectancy has increased significantly. Further advances in life expectancy are likely following recent progress toward the introduction of rational-design drugs and the development of vaccines targeting the most common lethal infectious diseases, such as malaria.

However, the comforting thought that any threats posed by infectious diseases are a thing of the past, is a perilous delusion. Climate change, large-scale human migration, increased medical utilization of antimicrobial agents and widespread use of nontherapeutic antimicrobial growth promoters are just a few factors that have led to a set of circumstances where infections now occur that are resistant to all current antibacterial options. This rising global health hazard is emphasized by the fact that infectious diseases are responsible for 25% of all human deaths worldwide. We are witnessing the re-emergence of diseases such as cholera, previously thought to be largely under control. Equally, there is a constant succession of newly identified infectious diseases caused by infectious agents that can jump from animals to humans, most recently exemplified by the appearance in 2013 of a new avian influenza strain termed H7N9, which is considered to be more easily transmissible from birds to humans then H5N1. The extraordinary ability of pathogens to change and adapt has resulted in the emergence of multidrug-resistant strains of widespread infectious diseases such as tuberculosis (TB), which are virtually untreatable and often fatal. Furthermore, the decreased immune response due to HIV infection has led to a resurgence of TB among millions in whom the disease has been dormant. HIV infection itself, first recognized in 1981, has caused a pandemic that is still in progress. The fragility of public health infrastructures is also characterized by the increasing incidence of nosocomial infections, exemplified by the emergence of Methicillin-resistant *Staphylococcus aureus* (MRSA) and diarrhea ascribed to *Clostridium difficile*. Taken together, these events emphasize the variability of infectious disease death rates, the uncertainty of disease emergence and the consequent continued threat to public health.

It is not surprising, then, that tremendous efforts are being made to improve the diagnostics of infectious diseases. The most important feature of an effective diagnostic assay is its capacity for the reliable identification of a targeted pathogen. At the same time, assays should be rapid, uncomplicated, and capable of being carried out in the field ("point-of-care"; POC). Unfortunately, the continued application of legacy assays entails that many "gold standard" diagnostic technologies are slow, expensive, require highly skilled personnel, and are not practical for POC applications. Hence, there is enormous pressure to implement ongoing molecular scientific and technological advances in the development of improved disease surveillance and control systems.

The combination of molecular diagnostics with therapeutics constitutes a key component of integrated healthcare and molecular testing methods have become the strategy of choice for the detection of many infectious diseases. They can provide sensitive and reliable results, detect infectious pathogens both qualitatively and quantitatively, and produce rapid test results enabling better and less expensive patient care. Although a qualitative measure of the infectious agent is often sufficient for the detection of many infectious diseases, there are several clinical conditions where the quantification of certain viruses is useful. This is true for measuring viral load in those patients immunosuppressed by HIV infection in order to monitor response to treatment. Molecular testing can provide clinicians with faster test results, often measured in minutes, compared with traditional microbiological methods. Combined with a move towards field-based, automated, simpler and higher throughput technologies, this development promises to generate a wealth of data establishing organism numbers and types as well as information about virulence and resistance determinants that influence disease severity. This will result in better patient care, translating into improved patient outcome, shortened hospital stays, and reduced cost of antimicrobial treatment.

5.1 Nucleic acid-based tests

Nucleic acid-based tests (NATs) encompass a range of amplification technologies and sophisticated detection methods that are readily adaptable for use in the diagnostic laboratory and have a number of key advantages for clinical testing, as compared with classical approaches (Tab. 5.1). A pathogen's nucleic acid provides the ultimate biomarker, as a carefully chosen target sequence will result in its exclusive identification. Alternatively, if a broader test is required, it is as easy to select less specific target sequences that classify groups of related pathogens. On the other hand, reliance on knowledge of the nucleic acid sequence introduces an important limitation of most NATs (next generation sequencing, NGS, is the exception). Assay design requires comprehensive information on sequences and sequence variations for targeted pathogens. In their absence, most NATs cannot detect unknown pathogens. However, the availability of genome sequences obtained from NGS has revolutionized the fields of microbiology and infectious diseases, resulting in the publication of more than 1,000 bacterial, 3,000 viral and nearly 100 fungal genomes, including representatives of all significant human pathogens. This permits the increasing penetration of NATs into pathogen diagnosis and genotyping as well as the detection of novel virulence and antibiotic resistance markers. These developments are entrenching NATs as the

Tab. 5.1: Advantages of nucleic acid-based tests.

Speed
Reduced risks of imprecise diagnosis, inappropriate therapy, over-use of antibiotics, development of drug resistance, reduced health-care costs
Sensitivity
No need for pre-enrichment
Specificity
Specific differentiation of closely related species or subtypes
Characterization of sequence variation
Antibiotic resistance gene mutations, thus aiding both diagnosis and treatment of infectious diseases
No need for growth in culture media
Identification of pathogens for which routine growth-based culture and microscopy methods are inadequate
Detection of pathogens in archival samples
Retrospective studies

---

benchmark for diagnosis of an ever-increasing range of pathogens, and their use is driving improvements in disease management whilst helping to decrease the costs associated with patient care.

NATs can detect pathogens using one of two amplification strategies: one relies on target amplification, the other on signal amplification. Both follow a basic, three-step workflow: (1) nucleic acid is made accessible from a biological sample; (2) a specific nucleic acid is amplified (target amplification) or targeted (signal amplification) by pathogen-specific DNA or RNA oligonucleotides; and (3) amplification products (target amplification) or targeted oligonucleotides (signal amplification) are detected.

In the following sections, methods of major importance for detection and/or monitoring of infectious agents in the routine diagnostic laboratory are described.

5.2 Target amplification methods

Target amplification methods are enzyme-mediated processes that use a single enzyme or multiple enzymes to increase the amount of target nucleic acid in a thermal or isothermal amplification reaction. They are analogous to finding a needle in a haystack by making a haystack out of the needle. This permits straightforward detection of the amplified targets, most conveniently using fluorescent reporter molecules. The main shortcoming of this strategy is the constant threat of contamination with product molecules that generate false positive results; however, this risk can be minimized through a range of precautions.

Amongst NATs, the polymerase chain reaction (PCR) continues to be the molecular enabling technology *par excellence*. Its diversification, enhancement and maturation have been instrumental in facilitating the emergence of molecular diagnostics, and today PCR technology is the most widely used tool for qualitative and quantitative detection of the small amount of infectious agent nucleic acid usually present in clinical specimens. Nevertheless, PCR is only one of several molecular diagnostic technologies and, despite its immense advantages, there are some practical disadvantages to its use. These relate to the requirement for specialized instrumentation, need for extensive assay validation and quality assurance and, not least, problems associated with licensing and intellectual property issues.

5.2.1 Real-time polymerase chain reaction (qPCR)

5.2.1.1 General considerations
qPCR is firmly established as the method of choice for the detection of nucleic acids. Today, in the routine diagnostic laboratory, molecular assays based on qPCR have largely replaced those based on conventional PCR. For the foreseeable future, it is likely that the power, simplicity, low cost and, not least, familiarity of qPCR will continue to make it the benchmark technology for molecular diagnostics. The features that make qPCR well suited to the clinical environment can be summarized as follows:
- Speed: assay reaction times are typically measured in minutes
- Convenience: homogenous assay, hence no need for post-amplification processing and reduced likelihood of contamination
- Simplicity: the assay requires two primers, an optional probe and a single enzyme
- Sensitivity: single copy targets can be detected, if not quantified
- Specificity: a well-designed assay is specific for a single target, but mismatch-tolerant assays are easily designed
- Robustness: a wide range of reaction conditions will yield results
- High throughput: thousands of assays can be carried out on a single run
- Quantification: typically over a huge dynamic range (eight to nine orders of magnitude)
- Familiarity: PCR has been around for more than 25 years and its advantages and disadvantages are well understood
- Cost: all assay reagents are inexpensive; together with the trend towards smaller reaction volumes the costs per assay can be very low

No other method in current use combines all these characteristics. The emergence of the swine flu pandemic in 2009, and the extraordinarily rapid appearance of qPCR-based assays for its detection and genotyping illustrates this very clearly. It is also worth noting, that reverse transcription (RT)-qPCR tests were shown to be significantly more sensitive than any of the influenza point-of-care tests currently on the market. A comparison with other technologies used in routine diagnostic pathology practice suggests that RT-qPCR is as reliable as legacy techniques, and is frequently more cost-effective and less time consuming. In addition, continuous improvements to reagents and enhancements of the technology itself are resulting in a stream of novel applications, including multiplex assays for the simultaneous detection of different targets and there is no doubt that the awareness of its diagnostic potential is becoming ever more widespread. Numerous qPCR and RT-qPCR assays have been developed to allow the identification of single viruses and bacteria. Broad-range PCR is a useful modification if the aim of the assay is the detection of pathogens in sterile site specimen such as blood or cerebrospinal fluid, described as "molecular petri dishes". Primers complementary to a conserved target region are used, such as the 16S rRNA bacterial gene or the 18S rRNA gene of fungi. Amplification products are sequenced and compared to the sequences held in web-based databases. The major drawback with this approach is that there is risk of amplifying DNA that may be contamination of the specimen or the PCR reagents themselves, especially the Taq DNA polymerases, resulting in false-positive results. Nevertheless, a comparison of various diagnostic methods regularly describes the performance advantages of qPCR-based assays.

The increasing number of assays, combined with the need to run replicates, controls and standards for quantification is accelerating the trend towards automation. This has the added advantage, that it improves the accuracy, robustness and reliability of qPCR diagnostic assays, since automation minimizes variation caused by pipetting, operators or between laboratories. Furthermore, use of robots is resulting in the development of protocols that require only very small reaction volumes, saving time (reactions can be carried out faster) and cost (fewer reagents are required).

There are several drawbacks to qPCR that are applicable to some of the other NAT methods:

1. The need to use purified nucleic acids limits the practicality of qPCR in a clinical setting, since it increases the time needed for sample preparation and introduces the likelihood of contamination. qPCR is also sensitive to environmental inhibitors that are concentrated along with pathogens during sample processing.

2. The small volume assayed may lead to false-negative results because of the low bacterial counts or virus titers.

3. The fact that assays determine only total pathogen number and do not provide information whether a pathogen has the ability to establish an infection or not. This last point is of particular significance, since the corollary is that a positive result may not necessarily pose a public health threat. Consequently, different and/or more complex combinations of assays are required that can determine viability, e.g. a combination of cell culture techniques followed by qPCR assays that detect replicating pathogen RNA present in the cell. However, this approach is not ideal as it requires additional analysis time, mRNA extraction, and qPCR reactions, increasing the potential for contamination.

5.2.1.2 qPCR detection formats

Detection formats include detection of double-stranded DNA and probe detection formats. Dyes such as LC Green, Eva Green, Chromofy, BEBO, and SYBR Green bind any double-stranded DNA and detection is monitored by measuring the increase in fluorescence throughout the cycle. A fluorescence signal is not only generated by amplification products, but also by primer dimers and other PCR artifacts. This technology is, thus, similar to detection of amplification products by gel electrophoresis and, although not sufficiently specific for use in the routine clinical laboratory, may be useful for a first evaluation during early developmental steps. However, recent developments in high resolution melting technology are generating new applications for these chemistries in the analysis of genetic variability of pathogens (see Section 5.2.1.3).

For primary detection in the routine clinical laboratory, use of probe-based assays is obligatory. To ensure strong binding of the probe(s) during annealing, the melting temperature of the probe(s) should be 5–10°C higher than that of the primers. Consequently, probes are usually between 25 and 30 nucleotides long and their design may be sometimes difficult to optimize because of formation of secondary structures or sequence variability. To overcome these problems, locked nucleic acid (LNA) probes may be used. LNA is a new class of DNA analog whose incorporation into oligonucleotides results in a significant increase in the thermal stability of duplexes with complementary DNA. A general increase of 3–8°C per modified base may be expected. Consequently, shorter probes can be used for hybridization-based assays.

Probe detection formats, which have been most frequently adapted to real-time instruments, include hybridization probes, hydrolysis probes, molecular beacons, and scorpions.

Hybridization probes

The hybridization probes format uses two different fluorescence-labeled oligonucleotides. The donor probe carries a fluorescein label at its 3′-end, whereas the acceptor probe is labeled with a different fluorescein label at its 5′-end. When the fluorescein at the donor probe is excited, it emits fluorescent light at a certain wavelength. The sequences of the two probes are selected so that they hybridize to the amplified DNA fragment in a head-to-tail arrangement, thereby bringing the two fluorescent dyes into close proximity. When both of the dyes are in close proximity, the energy emitted excites the dye attached to the acceptor probe that subsequently emits fluorescent light at a longer wavelength. This energy transfer, referred to as fluorescent resonance energy transfer (FRET), is highly dependent on close proximity (between one and five nucleotides) of the oligonucleotides. The increasing amount of measured fluorescence is proportional to the increasing amount of DNA generated during the ongoing PCR process. Because the signal is only emitted when both oligonucleotides are hybridized, fluorescence is measured just after the annealing step. After annealing, the temperature is raised and the hybridization probes are displaced by the polymerase. At the end of the elongation step, the amplification product is double-stranded and the probes are too far apart to allow FRET.

When using the hybridization probes format, it is essential that both of the probes cannot be elongated during PCR. For the donor probe, the fluorescein at the 3′-end blocks elongation, whereas, for the acceptor probe, a nonfluorescent blocking agent, typically a phosphate group is linked to the 3′-end. A sudden loss of probe fluorescence has been observed. Photobleaching and/or repeated freeze-thaw cycles have been correlated with the phenomenon, but recently an additional mechanism for FRET probe failure has been described: loss of the phosphate blocker from the 3′-end of the acceptor probe. If this occurs, then the acceptor probe can act as a PCR primer. To prevent this, the phosphate blocker may be replaced by a carbon spacer as a blocker of potentially enhanced stability.

Amplification products are designed to be not too long, because shorter amplification products amplify more efficiently than longer ones and are more tolerant to reaction conditions. For gene expression studies, the amplification product should be designed across an intron–exon boundary so that complementary regions in any contaminating genomic sequence are not amplified and products represent PCR of the cDNA template. If using FRET hybridization probes, use of LNA may be of special importance. Introduction of LNA residues offers the advantage of modulating the required melting temperature of probes without any modification of the sequences. This has been shown to be very useful for single nucleotide polymorphism detection.

Hydrolysis probes

The hydrolysis probe format uses an oligonucleotide with a fluorescent label (reporter dye) at its 5′-end and a quencher dye, which lacks native fluorescence to reduce background fluorescence, at its 3′-end. The hydrolysis probe anneals to the target DNA. During elongation, the 5′-exonuclease activity of the polymerase cleaves the probe, separating the reporter dye from the quencher. Consequently, emission of fluorescent light from the reporter dye can be detected during the data acquisition step. In contrast to all other detection formats, complete hydrolysis of the probes by the DNA polymerase is essential to yield precise results. In addition to choice of the adequate polymerase, factors that additionally influence hydrolysis include concentration of probes, the primer–probe distance, and avoidance of regions that would produce either primer:primerdimers or primer:probe dimers.

Molecular beacons

The hairpin-shaped molecular beacon is nonfluorescent because of its stem hybrid that keeps the reporter dye close to the quencher. When the probe sequence in the loop hybridizes to its target, the quencher is physically separated from the reporter dye and fluorescence is restored. Assays that use molecular beacons are more difficult to optimize than those using hydrolysis probes. The stem structure forms by an intramolecular hybridization event and the signal yield is very sensitive to hybridization conditions. Calculation of the assay kinetics may thus be rather complicated.

Scorpions

The scorpion consists of a specific probe sequence that is held in a hairpin loop configuration by complementary stem sequences on the 5′ and 3′ sides of the probe. The fluorophore attached to the 5′-end is quenched by a moiety joined to the 3′-end of the loop. The hairpin loop is linked to the 5′-end of a primer via a PCR stopper. After extension of the primer during PCR amplification, the specific probe sequence is able to bind to its complement within the same strand of DNA. This hybridization event opens the hairpin loop so that fluorescence is no longer quenched and an increase in signal is observed. The PCR stopper prevents read-through, which could lead to opening of the hairpin loop in the absence of the specific target sequence.

5.2.1.3 Melting curve analysis

The temperature at which double-stranded DNA separates can vary greatly depending on the length, but mainly sequence (GC content). After completion of the amplification protocol, the temperature is steadily increased while the fluorescence is monitored. Fluorescence decreases as the temperature increases. At a certain temperature, an abrupt decrease of fluorescence can be observed because of the melting of a product. The melting temperature of a product is defined as the temperature at which half of the DNA is single stranded, and represents the steepest decrease of the fluorescent signal. This can be identified conveniently as the peak value in the negative derivative of the melting curve. Melting curve analysis can be performed for all detection formats, except for the TaqMan probe format, because signal generation depends on the hydrolysis of the probe.

The melting temperature profile can provide additional information, e.g. the genotype of the DNA product. Even single-base differences in heterozygous DNA can change its melting temperature. A heterozygous sample contains two DNA sequences, each of which melts at a different temperature, resulting in a two-peak curve. Unexpected melting peaks may indicate primer:primer or primer:probe dimers or sequence variants (Fig. 5.1).

The recent introduction of high-resolution melting (HRM) analysis has hugely boosted the potential of this technology as a clinical diagnostic tool. HRM uses a combination of new-generation DNA dyes with improved saturation properties, high-end instrumentation and sophisticated analysis software and is homogeneous, closed-tube, rapid (1–5 min), and non-destructive. It vastly increases the power of classical melting curve analysis by allowing detailed analysis of the thermal denaturation profiles of a double-stranded DNA. The most important HRM applications for infectious diseases are genotyping, variation scanning and mutation analysis of bacterial,

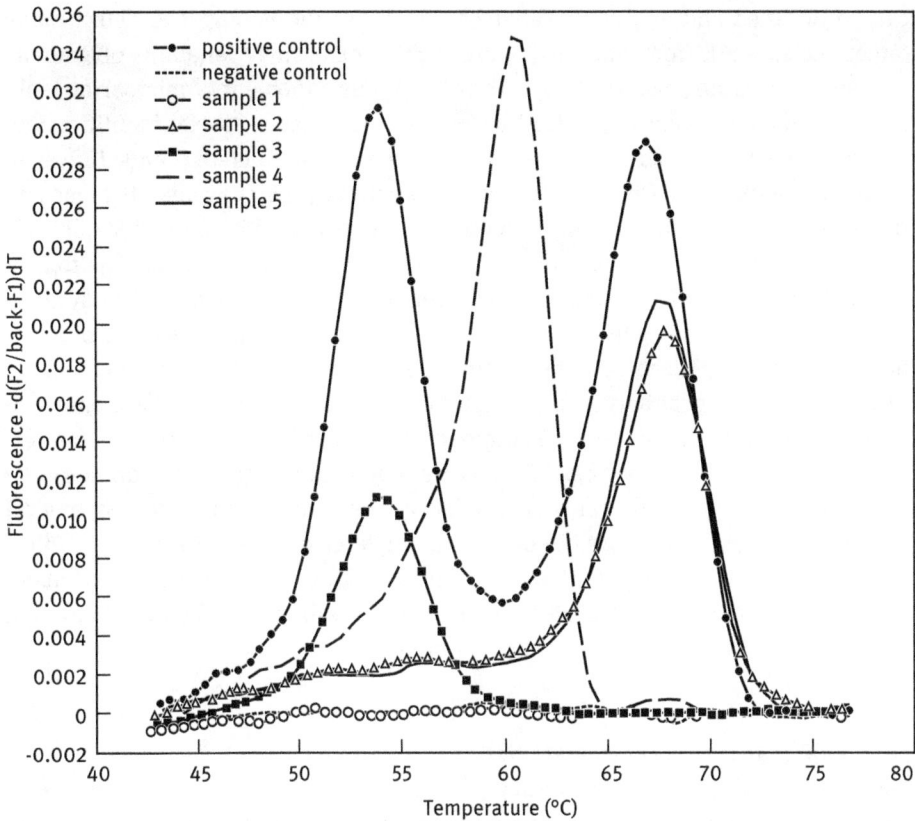

Fig. 5.1: Detection of herpes simplex virus (HSV) type 1 and type 2 DNA by qPCR. Melting curves of clinical samples. HSV-1 was detected in sample 3; HSV-2 in samples 2 and 5. The positive control contains both, HSV-1 and HSV-2. Sample 4 shows an unexpected melting peak indicating sequence variation.

viral, and parasitic pathogens. It has similar or superior sensitivity and specificity compared to methods that require physical separation. With high resolution melting, single base changes such can be discriminated reliably without probes and simultaneous genotyping or scanning of an entire amplification product can be performed at the same time in the same tube, vastly decreasing or eliminating the need for re-sequencing in genetic analysis. Recent uses for HRM include the rapid and reliable monitoring of HIV diversity in patients with different stages of HIV, identifying antibiotic resistance in bacterial pathogens and discrimination of fungal pathogens.

5.2.1.4 Microarrays

Microarrays are generally composed of thousands of specific probes spotted onto a solid surface. Each probe is complementary to a specific nucleic acid sequence and hybridization with a labeled complementary sequence results in a signal that can be detected and analyzed. This technology has the potential to detect many target nucleic acids from multiple organisms; the increasing availability of genomic sequences of pathogens, together with the rapid development of microarray technology, make it a potentially powerful tool for simultaneous species identification, detection of virulence factors and antimicrobial resistance determinants. However, wide scale adoption of DNA microarrays is hindered by their low analytical measuring range and variable hybridization efficiencies, resulting in unreliable quantification and false-positive/negative results. There may also be problems due to lack of sensitivity, as multiplex PCR precedes microarrays. Nevertheless, test systems including the microarray technology are on the market because the proof of a sufficient sensitivity of each pathogen included in such an assay is not obligatory for IVD/CE-labeling. Sensitivity is important, because in some instances it is essential to detect any pathogen at the earliest opportunity to start with pre-emptive therapy. For example, when monitoring children after bone marrow transplantation due to hematologic disease, pathogen load will be very low and reliable detection of viral reactivation is currently possible only with single or duplex qPCR assays. Unfortunately, these problems are inherent to this technology; practical applications of microarray technology as clinical diagnostic assays will require the introduction of new detection chemistries and analysis approaches.

5.2.1.5 qPCR technical considerations

The most straightforward application of qPCR relates to its use for the detection and discrimination of DNA. This is because sample handling, template preparation, assay protocols, and data interpretation are all relatively straightforward. The increased availability of a wide range of commercial assays has also reduced the need to generate "home-brew" assays (laboratory-developed assays), with all the associated problems of assay design in general and primer design in particular resulting in poor amplification efficiencies, which have a direct effect on the assays sensitivity and, by affecting reproducibility, will also adversely affect assay specificity.

Nevertheless, inhibition of the PCR assay by factors co-purified during nucleic acid extraction can be an important factor contributing to poor assay performance. It is well worth testing every sample for the presence of inhibitors that can affect different targets, enzymes, and assays in different ways. The question of inter-laboratory variation has also not been solved, with reports continuing to refer to this issue. Finally, improper data analysis can make a positive result appear as negative and vice versa. Hence, it is essential that the standard operating procedure includes a detailed description of the parameters used for data analysis.

When targeting RNA, the most common causes for technical inadequacy of NATs in general are inconsistent sample handling, RNA isolation, RNA quality control and, for RT-qPCR assays in particular, the RT step. Crucially, the accuracy of pathogen detection is highly dependent on RNA quality, both in terms of its integrity as well as in terms of the lack of inhibitors co-purified during the extraction procedures.

The conversion of RNA to cDNA is a highly variable step and RT-qPCR results vary considerably with the choice of reverse transcriptase. The choice of primer location on the target RNA also can yield significantly different results, as RNA adopts a tight secondary structure characterized by extensive intra-strand base pairing resulting in stem-loop structures. If RT primers are designed to target stems, rather than loops, or if the amplification product can adopt secondary structures, the efficiency of the RT step can be significantly compromised. Characteristically, this results in non-quantitative and non-reproducible results. Finally, it is essential that the specificity of primers is confirmed prior to developing qPCR assays to avoid false-positive results caused by non-specific priming.

One of the most shocking examples of the enormous implications for the health and lives of individuals that result from inappropriate use of this technology is provided by the use of RT-qPCR data that appeared to demonstrate the presence of measles virus RNA in children with developmental disorders. It provided sustenance to the controversy surrounding the triple measles mumps and rubella (MMR) virus vaccine, as the data were interpreted as providing hard evidence for a link between MMR, gut pathology and autism. However, a detailed analysis of the raw data underlying that report carried out by one of the authors acting as an expert witness to the UK High Court and the US Vaccine Court, revealed that these data were obtained amongst a catalogue of mistakes, inaccuracies and inappropriate analysis methods as well as contamination and poor assay performance. A reanalysis of the data concluded that the assay had been detecting DNA and since measles virus is an RNA-only virus, the RT-qPCR data had been erroneously interpreted. A recent paper was unable to reproduce the original findings and concluded that there was no link between autism and enteropathy and the US Vaccine court has now ruled that there is no link between the MMR vaccine and autism.

The more widespread acknowledgement of these problems has resulted in an initiative aimed at improving the reliability of qPCR data. Specifically, guidelines have been published that provide a blueprint for good assay design and draw up specifications for the minimum information for the publication of quantitative PCR experiments (MIQE). MIQE is modeled on similar guidelines drawn up for DNA microarray analysis, proteomics experiments, genome sequence specification and those under discussion for RNAi work and metabolomics, initiatives coordinated under the umbrella of MIBBI, Minimum Information for Biological and Biomedical Investigations (www.mibbi.org). These guidelines will improve the relevance, accuracy, correct interpretation and repeatability of qPCR data, and thus help ensure the integrity of the scientific literature, promote consistency between laboratories and increase experimental transparency. The publication of the MIQE guidelines has turned out to be a defining event in the maturation of qPCR technology. There has been a universally positive response from instrument and reagent manufacturers, extensive publicity in print, online, and at scientific meetings and within three years of publication there are more than 600 citations in the peer-reviewed literature.

5.2.1.6 Digital PCR

Accurate quantification by qPCR can be limited due to the requirement for assay calibration with standards whose performance characteristics must be equivalent to the samples being assessed (see Chapter 3). Furthermore, qPCR assay sensitivity for the detection of mutations in genomic DNA is around 1% at best, which is not sufficient for consistent detection of minor allele frequencies; in addition, the limit of quantification between two samples is approximately two-fold, which is generally not sufficient for distinguishing copy number variants with heterogeneous material. These limitations are addressed by digital PCR (dPCR) measurement, which uses a combination of limiting dilution, end-point PCR, and Poisson statistics to obtain an absolute measure of nucleic acid concentration. dPCR is set up like qPCR and carried out on a single sample using the same primers and probes. Unlike qPCR however, the sample and dPCR assay mixture are partitioned into thousands or millions of discrete uniform small volume reactions, such that there is either no target or one target molecule present in any individual reaction chamber. The PCR reaction is carried out as an endpoint reaction (real-time is possible in some systems) and compartments containing a target will generate a fluorescent signal whereas sections with no target will remain dark. A positive reaction is recorded as "1", whereas a reaction with no target is counted as "0". This binary nature of the results is the reason this PCR variant is called "digital" PCR.

If the sample is sufficiently dilute, each of these positive reaction chambers can be assumed to have contained only a single target molecule. Hence the total number of "positive" reactions describes the number of original target molecules in the entire volume, giving an "absolute" target copy number. This way of recording results explains the description of this technology as digital PCR. Furthermore, the concentration of the target is easily determined by dividing that number by the total measured volume (the total number of reactions multiplied by the individual reaction volumes).

If the sample has not been diluted sufficiently, some of the positive signals will have originated from chambers containing two or more target molecules. Consequently, simply counting positives will underestimate the actual number of molecules. This problem is corrected by using the Poisson equation $A = -\log_e(1-P)$, which calculates the average number of molecules per aliquot (A) from the observed proportion of positive aliquots (P). Alternatively, relative target copy numbers can be determined by comparing its abundance to that of a reference sequence, for example a gene known to be present in two copies per diploid cell.

Precision and sensitivity of dPCR depend on the total number of reactions that are carried out and increase as more aliquots are analyzed. In general, dPCR is more accurate and less ambiguous than qPCR, as measurement uncertainty derives only from error in the measured volume or the presence of more than a single target molecule in a compartment. Hence dPCR can measure differences as small as 1.2-fold (±10%) in mRNA level and identify alleles occurring with frequencies of less than 0.01%. Whereas qPCR will amplify both abundant and rare alleles present in a single sample, with the signal from the more abundant sequences dominating and masking the signal from rare mutants, dPCR overcomes this difficulty inherent in amplifying rare sequences. The efficiency of the PCR reaction is less critical than it is for qPCR, where small differences can result in significantly different crossing points and hence quantification measurements. In contrast, PCR efficiency for dPCR must be sufficient only to generate a fluorescent signal at the end of the reaction. On the other hand, its analytical measuring range is much narrower than that of qPCR and whereas qPCR can be carried out to detect different targets per sample run, this is not so with dPCR. Nevertheless, there are many applications that will benefit from the power and accuracy of dPCR. In addition to rare allele detection in oncology these include non-invasive prenatal diagnostics, measurement of copy number variation in heterogeneous samples, single cell gene expression, and validation of next-generation sequencing. dPCR can also be used for pathogen detection, where the method enables high precision detection and quantification of rare pathogen genomic sequences from a background of cellular DNA or RNA.

5.2.2 Isothermal amplification techniques

5.2.2.1 Nucleic acid sequence-based amplification (NASBA) and transcription mediated amplification (TMA)

Nucleic acid sequence-based amplification (NASBA), also known as self-sustained sequence replication (3SR), and transcription mediated amplification (TMA) are iso-thermal RNA amplification methods that mimic retroviral replication. Both methods use a combination of multiple enzymes: NASBA uses three, the avian myeloblastosis virus (AMV) reverse transcriptase/DNApolymerase, T7 RNA polymerase and a separate RNase H, whereas TMA makes use of the intrinsic RNaseH activity of the reverse transcriptase. Both methods can target either DNA, albeit less efficiently, or RNA. The polymerization process generates RNA amplification products; these have the theoretical advantage of reducing the possibility of carry-over contamination since RNA is more labile than DNA. The methods use a primer that contains target-specific as well as T7 RNA polymerase promoter sequences. Following hybridization of this primer to the target RNA, reverse transcriptase creates a cDNA copy of the target RNA, the resulting RNA:DNA duplex is degraded by RNase activity and a second primer binds to the cDNA. A new strand of DNA is synthesized from the end of this primer by reverse transcriptase, creating a double-stranded DNA molecule with a T7 RNA polymerase promoter. RNA polymerase recognizes its promoter sequence and initiates transcription. Each of the newly synthesized RNA amplification products re-enters the amplification process and serves as a template for a new round of replication leading to an exponential expansion of the RNA amplification products. A target-specific probe using a real-time format detects the amplification products generated in these reactions.

For detection of infectious agents, the NASBA technology plays only a minor role. Few laboratories utilize this assay for detection of human immunodeficiency virus type 1 (HIV-1) and even fewer for some other viruses including enterovirus and herpes simplex virus. Recently, a NASBA assay which detects E6/E7 mRNA from the high risk human papilloma virus types 16, 18, 31 and 45 was brought to the market and is utilized in a few specialized laboratories. Several laboratories employ the TMA technology for qualitative detection of *Chlamydia trachomatis* and *Neisseria gonorrhoeae*. Furthermore, this assay is utilized for qualitative detection of hepatitis C virus (HCV) and HIV-1 in several transfusion medicine facilities.

5.2.2.2 Strand displacement amplification (SDA), Loop-mediated isothermal amplification (LAMP), and Q-LAMP

Strand displacement amplification (SDA) is a somewhat complex, isothermal amplification procedure. Amplification consists of two steps: a target generation step followed by an exponential amplification phase that replicates the target sequence through a series of complex extension, nicking, and strand displacement steps. The products contain a detector probe-annealing region and amplification product detection occurs simultaneously with amplification. The detector probe consists of a target specific hybridization region at the 3′ end and a hairpin structure at the 5′ end. The loop of the hairpin contains a restriction enzyme recognition sequence. The 5′ base is conjugated to a fluorophore donor molecule, while the 3′ base of the hairpin stem is conjugated to an acceptor molecule. In its native state, the hairpin maintains the donor and the acceptor molecules in close proximity. When the donor is excited, the fluorescent energy is transferred to the acceptor molecule, and little fluorescence is observed. As the hairpin anneals to the target, another complex set of displacement, extension, and restriction steps frees the donor from the quenching effects of the acceptor, and allows fluorescence to be observed. Today, the SDA assay is mainly used for detection of *Chlamydia trachomatis* and *Neisseria gonorrhoeae* in the routine diagnostic laboratory.

Conventional loop-mediated isothermal amplification (LAMP, Eiken Chemical Co. Ltd) amplifies nucleic acid under isothermal conditions. This method employs a set of four primers that bind to six distinct sequences on the target nucleic acid, resulting in high specificity. An inner primer initiates LAMP and subsequent strand-displacement DNA synthesis, primed by an outer primer, releases a single-stranded DNA. This acts as a template for DNA synthesis primed by the second inner and outer primers. These hybridize to the opposite end of the target sequence, producing a stem-loop DNA structure. Subsequently, as LAMP progresses, an inner primer hybridizes to the loop on the product and initiates displacement DNA synthesis, yielding the original stem-loop DNA and a new stem-loop DNA that is twice as long. This is a rapid and stable isothermal process, producing up to 10^9 copies of the target in less than one hour. The final stem-loop DNA products contain multiple loops formed by the annealing of alternately inverted repeats of the target in the same strand.

LAMP technology has been licensed and further developed to produce a quantitative (fluorescence-quenching) loop-mediated isothermal amplification method, Q-LAMP (DiaSorin S.p.A.). Q-LAMP is a rapid real-time fluorescent technique that can be used for multiplexed applications, allowing the quantitative or qualitative analysis of single or multiple nucleic acid targets in a single reaction. Like LAMP, Q-LAMP is based on the recognition of multiple primer binding regions on the target nucleic acid and amplification of the target sequence, facilitated by a polymerase with strand displacement activity. Quantification is achieved through the use of fluorophore-labeled primers and known calibrators within the reaction solution, which fluoresce in their free form but are quenched when bound. As the fluorophore-labeled primers are consumed during amplification of the target sequence, there is an observed decrease in fluorescence. This produces a characteristic real-time quenching profile when measured over time. Notably, unlike PCR, this method achieves single tube RNA amplification without the need for an additional RT step. The speed and simplicity of Q-LAMP is of clinical value in a number of infectious disease and onco-hematology applications.

5.2.2.3 Recombinase polymerase amplification (RPA)
Recombinase polymerase amplification (RPA) is a multienzyme isothermal amplification technique that makes use of the DNA binding and unwinding properties of prokaryotic recombinases. Target-specific primers and a recombinase are added to a sample and the recombinase/oligonucleotides complex scans for complementary sequences within target DNA samples. If the target sequence is present, the recombinase displaces one DNA strand with the oligonucleotides that serve as primers for the initiation of DNA synthesis by a DNA polymerase. The assay also contains single-stranded DNA-binding proteins (SSBs) that attach to and stabilize the displaced strand. These interactions result in the duplication of the original template from primer-defined points in the DNA sample. Repetition of this process results in exponential DNA amplification.

For maximum portability, assays can be performed in a real-time fluorometer running on batteries and able to monitor up to eight RPA reactions. RPA assays can also be used in combination with alternative, commonly used detection techniques, such as gel electrophoresis, dipsticks and standard fluorescence plate readers and qPCR machines (with a DNA binding dye or an additional component, exonuclease III, which will degrade the hybridization probe and generate a real-time read-out). The assay's main advantages are

- True isothermal protocols: There is no requirement to melt double-stranded DNA or to unfold RNA hence RPA is a true one-step amplification procedure, distinguish-ing it from other isothermal amplification processes. RPA operates from ambient temperatures to as high as 45°C and tolerates fluctuations within these limits. It is not affected by inadvertent off-temperatures or low temperature set-up. The assay works at typical ambient temperatures, albeit more slowly. The entire reaction system is stable as a dried formulation and can be transported safely without refrigeration.
- Robustness: RPA is significantly more resistant to crude samples than PCR, hence can be used on crude samples, including blood and nasal swabs, without any requirement for nucleic acid purification making it ideally suitable for field use and point-of-care applications.
- Speed: Detection occurs within 10–15 min. Combining this performance with simple rapid sample preparation and decentralized point-of-use, RPA enables diag-nosis within a half-hour or less compared to the 24-hour turnarounds cur-rently typical for clinical samples requiring analysis at centralized facilities.

In addition, RPA is as sensitive and specific as PCR-based assays and can be used as a multiplex assay to detect several targets simultaneously in one reaction.

These features make RPA a prime candidate for consideration as a convenient, portable diagnostics procedure. One recent practical application is for the digital quantitative detection of methicillin-resistant *Staphylococcus aureus* (MRSA) genomic DNA using a microfluidic digital RPA chip, which allows the simultaneous perfor-mance of over 1,000 nanoliter-scale RPA reactions. Most recently, a lab-on-a-chip platform was launched that can detect as few as 10 antibiotic resistant *Staphylococ-cus aureus* in less than 20 min, outperforming existing PCR based lab-on-a-chip plat-forms in terms of energy efficiency and time-to-result.

5.2.3 Next generation sequencing (NGS)

Next-Generation Sequencing (NGS) technologies have become indispensable tools in infectious disease research, including drug discovery and diagnostics. NGS involves extraction of DNA or RNA from the sample of interest, its conversion to an NGS library and the massively parallel sequencing of clonally amplified or single DNA molecu-les that are spatially separated in a flow cell. Sequencing is performed by repeated cycles of polymerase-mediated nucleotide extensions or by iterative cycles of oligo-nucleotide ligation, generating hundreds of megabases to gigabases of nucleotide sequence output in a single instrument run, depending on the platform. The arrival of improved chemistries and instrumentation has dramatically decreased sequencing cost and time and fuelled the demand for fast, inexpensive, and accurate genome information. NGS has a tremendous impact on the study of microbial diversity in environmental and clinical samples and has created the field of shotgun metagenomics, which targets the meta-genomes of entire communities rather than just the 16S rRNA of bacteria. This involves aligning sequence outputs to reference sequences and allows closely related species to be discerned and more distantly related species to be inferred. In addition, *de novo* assembly of the data set can identify potentially new species. Qualitative genomic information is obtained and analysis of the relative abundance of the sequence reads can be used to derive quantitative information on individual microbial species.

In practice, NGS has allowed the identification of unknown etiologic agents in human diseases, for example the discovery of never-before-seen pathogens responsible for the deaths of transplant patients. Its use has resulted in detailed studies of the viral and microbial species that occupy surfaces of the human body. Furthermore, it permits the identification of resistant clones constituting only a minor fraction of a pathogen population during the early stages of an infection and allows the analysis of bacterial genome patterns resulting in the evolution of pathogenicity.

Issues such as complexity of technical procedures, robustness, accuracy, cost, and bioinformatic bottlenecks are being addressed by the introduction of low-cost personal sequencing instrumentation. Further process streamlining, chemistry refinements and improved data handling will continue to drive cost and time reductions.

5.3 Signal amplification methods

Current signal amplification methods use a dual strategy to maximize analytical sensitivity: (1) optimized hybridization of label molecules to the target nucleic acid; and (2) generation of a profusion of signaling molecules. Amplification can involve label detection by enzymatic or non-enzymatic technologies; non-enzymatic technologies are not affected by the presence of enzyme inhibitors usually found in the clinical specimen, allowing simplified nucleic acid extraction procedures. Target detection can be enhanced using multiple enzymes, multiple probes, layers of probes, and reduction in background noise. The use of several probes or larger probes makes these assays less susceptible to target sequence heterogeneity.

There are advantages to detecting a target sequence directly. Most notably, a method that does not amplify the signal exponentially is more amenable to quantitative analysis. Even if the signal is enhanced by attaching multiple dyes to a single oligonucleotide, the correlation between the final signal intensity and amount of target is direct. Furthermore, there is no possibility of assay contamination by target amplification products, although accidental contamination of a sample with amplified reporter probes will result in false-positive results. As a rule, although sensitive, signal amplification is less sensitive than target amplification. In addition, the hybridization requirement makes these methods less rapid.

Signal amplification methods include the branched DNA (bDNA), the hybrid capture, the Invader and the ramification amplification assays. Today, signal amplification methods have largely been replaced by target amplification methods in the routine diagnostic laboratory with only the branched DNA and the hybrid capture assays being of some importance.

5.3.1 Branched DNA (bDNA)

The bDNA assay is a sandwich nucleic acid hybridization method that detects the presence of specific nucleic acids by measuring the signal generated by specific hybridization of branched, labeled DNA probes on an immobilized target nucleic acid. Signal amplification is achieved by sequential or simultaneous hybridization of multiple oligonucleotides, assembling a branched complex structure on the immobilized target nucleic acid. In general, one end of bDNA binds to a specific target and the other end has many branches of DNA. The branches contain several hundred labels and so amplify detection signals. The bDNA assay has advantages: (1) as it works directly from cell lysates or tissue homogenates, there is no need for nucleic acid isolation; (2) as it avoids potential bias from the enzymatic conversion of RNA into DNA and any subsequent PCR; and (3) as it can be multiplexed, allowing the simultaneous detection of more than 30 targets. Furthermore, for specialized applications such as the detection of RNA targets from archival, formalin-fixed material, this method can result in equivalent or superior performance characteristics compared with methods requiring an RT step. The technology's main limitations are that it currently cannot reliably detect fewer than approximately 100 copies of target and that its dynamic range is limited to around four orders of magnitude. Because of the lack of sensitivity, the bDNA assay must not be employed for detection of pathogens in the routine diagnostic lab; however, may be used for monitoring purposes.

5.3.2 Hybrid capture assay

The hybrid capture assay detects and quantifies pathogens using solution hybridization, immunocapture and chemiluminescent signal detection steps. It uses long RNA probes for the detection of pathogen DNA, thus minimizing the risk of false negatives caused by sequence variation. RNA probes are allowed to hybridize with pathogen DNA and the hybrids are captured using an RNA/DNA-specific antibody attached to a solid phase support. Captured hybrids are detected with multiple antibodies specific for a particular RNA:DNA hybrid conjugated to alkaline phosphatase. Cleavage of a dioxetane substrate by the enzyme results in chemiluminescence that, when amplified, can be measured. Hybrid capture assays appear to lack sensitivity, results are more variable, and they may be more prone to contamination. Nevertheless, many routine diagnostic laboratories employ this assay for the detection of human papillomavirus.

5.4 What are the key challenges for the future?

Advances in second- and third generation massive parallel sequencing have been such that instrumentation and reagents are becoming affordable as well as fast. The huge advantage of such systems is that the millions of nucleotide reads per sample result in an almost complete spectrum of nucleotide information without prior knowledge or use of genetic information in advance. This makes them a valuable tool for diagnosing emerging infectious diseases. Combined with recently developed metagenomic methods, both techniques are proving to be powerful tools for simultaneous and assumption-free pathogen detection. Hence, healthcare diagnostics and biotechnology companies are teaming up aiming to drive the development of next-generation sequencing for the introduction of rapid, accurate sequencing-based infectious disease assays for the clinical diagnostics market.

This continued discovery and validation of new genetic markers of current, novel, and evolving disease-causing agents means that molecular methods will allow increasingly accurate and sensitive detection of an expanding range of known "alien" biological material in the body. Consequently, it will bring us ever closer to the ultimate goals: to characterize pathogens comprehensively, early, unequivocally and sensitively and so contribute to early, more effective treatment and to the monitoring of a patient's progress.

Future molecular diagnostic tests are likely to be simple, robust, and highly portable. Molecular dipsticks provide one more immediate solution for simple detection, although they still require an (ideally isothermal) amplification reaction to be carried out separately. In the medium term, the betting must be on miniaturized hand-held devices capable of delivering point-of-care tests in the absence of facilities such as electricity. The problem lies in the dichotomy inherent in requiring such a device to be as simple as possible, yet be capable of distributing a patient's nucleic acid test sample among the array of reactors and microvalves necessary to carry out the test. An interesting report details the development of a primer pair pre-loaded PCR array chip, in which the loading of the PCR mixture into an array of reactors and subsequent sealing of the reactors are realized by a novel capillary-based microfluidics device with a manual two-step pipetting operation. The chip is capable of performing parallel analyses of multiple gene targets and its performance was tested by amplifying 12 different gene targets and represents a step towards a future diagnostic lab-on-a-chip.

Another exciting development, especially in relation to its potential for automation and miniaturizing fluidic operation, involves a combination of gold nanoparticles conjugated with oligonucleotide probes, a sequential flow injection analysis system, beads injection technology and, less future-proof, real-time immuno-PCR.

It is also likely that the range of specimens available for molecular diagnostics tests will increase to include samples other than tissue, blood, or sputum. Urine has many advantages, not least those of accessibility and reduced extraction costs. DNA fragments, termed transrenal DNA from viruses and bacteria have been identified in the soluble portion of urine, which has tended to be discarded. However, upstream processes such as storage and extraction of urine have significant effects on DNA yield, resulting in variability in sample consistency and sensitivity of pathogen detection; hence there is a need to develop robust and inexpensive nucleic acid extraction methods from urine. Whilst current assays employ PCR-based technology, it is it not hard to envisage the application of even more sensitive, more practical technologies that will result in a simple urine test to diagnose rapidly infectious diseases at the point-of-care.

5.5 Take-home messages

- NAT methods represent an important pillar for the future of molecular diagnostics. This technology has opened up new detection possibilities and is becoming the new gold standard for the molecular detection of pathogens.
- Currently, qPCR provides the best compromise between ease of use, sensitivity, reliability and specificity of any molecular diagnostic assay. The main issues remaining are those concerning the potential for observing false-positive or negative results, problems detecting live pathogens, nucleic acid quality control issues, the requirement for sequence information of the pathogens being targeted and problems regarding the introduction of genetic alterations. Therefore, it cannot be over emphasized that there remains a constant need for assay validation to ascertain the reliability and reproducibility of established, modified, or newly developed technologies for routine analysis.
- dPCR provides an alternative for absolute quantification of nucleic acids and provides the potential for extremely accurate measurement of pathogen load.
- The development of lab-on-a-chip devices combining miniaturized reactors, advanced microfluidics, and sophisticated detection technologies will ultimately become the standard diagnostic tool-kit. It will detect nucleic acids; will it involve PCR?

5.6 Further reading

[1] Bustin, S.A., Nolan, T. (2004) Pitfalls of quantitative real-time reverse-transcription polymerase chain reaction. J. Biomol. Technol. 15:155–166.

[2] Bustin, S.A., Mueller, R. (2005) Real-time reverse transcription PCR (qRT-PCR) and its potential use in clinical diagnosis. Clin. Sci. (Lond.) 109:365–379.

[3] Bustin, S.A. (2008) Molecular medicine, gene-expression profiling and molecular diagnostics: putting the cart before the horse. Biomark. Med. 2:201–207.

[4] Bustin, S.A. (2008) Real-time quantitative PCR-opportunities and pitfalls. Eur. Pharm. Rev. 4:18–23.

[5] Bustin, S.A. (2008) RT-qPCR and molecular diagnostics: no evidence for measles virus in the GI tract of autistic children. Eur. Pharm. Rev. Dig. 1:11–16.

[6] Bustin, S.A., Benes, V., Garson, J.A., Hellemans, J., Huggett, J., Kubista, M., Mueller, R., Nolan, T., Pfaffl, M.W., Shipley, G.L., Vandesompele, J., Wittwer, C.T. (2009) The MIQE Guidelines: minimum information for publication of quantitative real-time PCR experiments. Clin. Chem. 55:609–620.

[7] Garson, J.A., Huggett, J.F., Bustin, S.A., Pfaffl, M.W., Benes, V., Vandesompele, J., Shipley, G.L. (2009) Unreliable real-time PCR analysis of human endogenous retrovirus-W (HERV-W) RNA expression and DNA copy number in multiple sclerosis. AIDS Res. Hum. Retroviruses 25:377–378.

[8] Kessler, H.H. (2007) Design and work-up of a new molecular diagnostic assay based on real-time PCR. In: Protocols for Nucleic Acid Analysis by Nonradioactive Probes, 2nd edn, Hilario, E., Mackay, J. eds., pp. 227–236.

[9] Lyon, E., Wittwer, C.T. (2009) LightCycler technology in molecular diagnostics. J. Mol. Diagn. 11:93–101.

[10] Murphy, J., Bustin, S.A. (2009) Reliability of real-time reverse-transcription PCR in clinical diagnostics: gold standard or substandard? Expert Rev. Mol. Diagn. 9:187–197.

[11] Baker, M. (2012) Digital PCR hits its stride. Nat. Methods 9:541–544.

6 Interpreting and reporting molecular diagnostic tests

Ranjini Valiathan and Deshratn Asthana

Today, molecular methods have become an essential part of the clinical microbiology laboratory, particularly for the detection and characterization of viral infections and for the diagnosis of diseases due to fastidious bacteria. The advantages of rapid turn-around time and high sensitivity and specificity are appealing. With the introduction of automation for DNA or RNA extraction followed by real-time PCR (qPCR), molecular laboratories have become more efficient and cost-effective. Even though qualitative and quantitative molecular technology can provide uniquely valuable information about patient conditions, clinicians may not gain maximal benefit from test results without clear and adequate details about test method and limitations and in-depth interpretation.

6.1 Detection of viral infections

Molecular testing methods allow laboratories and healthcare professionals to detect the presence of viruses causing infectious disease both qualitatively and quantitatively. A qualitative measure of the infectious agent is often sufficient for the detection of many infectious diseases. There are several clinical conditions where the quantification of certain viruses is required. Specimens suited ideally for viral quantification include EDTA whole blood, plasma, serum, CSF, and urine. However, specimens such as swabs, secretions, lavages, seminal fluids, stool and biopsies should be preferably used for qualitative detection of viruses.

Of particular importance for immunosuppressed patients, viral load can be determined using quantitative molecular tests to identify the amount of certain viruses such as cytomegalovirus (CMV), Epstein-Barr virus (EBV) and polyoma virus BK (BKV). Many people have been infected with these viruses and present with latent infection with low levels of viral DNA. Unless a patient becomes immunosuppressed, such as after a transplant or due to an immunosuppressive disease, these viruses tend to be clinically insignificant. The virus can reactivate in an immunosuppressed patient with dire consequences. In immunosuppressed patients, healthcare providers are able to monitor the virus quantitatively using molecular testing to determine when the viral load rises to critical levels. Furthermore, quantitative molecular tests are used to monitor the response to therapy against certain viruses such as hepatitis B virus (HBV), hepatitis C virus (HCV) and human immunodeficiency virus (HIV). Quantitative detection of infectious agents can be of particular importance when measuring tumor burden. For example, if EBV is present within a tumor, clinicians can gather important information concerning the size of a tumor by measuring the amount of EBV present in a patient sample using a quantitative molecular test.

6.2 Detection of bacterial infections

Unlike diagnostic virology, which detects target DNA as well as target RNA, most molecular diagnostic assays in medical bacteriology have been based around the amplification of DNA in a target gene rather than mRNA or other nucleic acid signal. This may be particularly important, as DNA is an extremely stable molecule, as opposed to mRNA, which has a short half-life. Generally, molecular diagnostics in clinical bacteriology is not concerned with regard to the viable status of an organism being detected, but is concerned with the qualitative detection of an organism in a symptomatic patient with the relative clinical presentation. Techniques such as PCR, nested PCR, multiplex PCR, qPCR, fingerprint methods such as Random Amplified Polymorphic DNA (RAPD), Amplified Fragment Length Polymorphism (AFLP), Restriction Fragment Length Polymorphism (RFLP), BOX-A1R-based repetitive extragenic palindromic-PCR, sequencing, Multilocus Sequence Typing (MLST), microarrays, and oligonucleotide probes are used for the detection of bacteria.

Amplification of DNA encoding ribosomal RNA (rRNA) genes in conjunction with DNA sequencing of the amplification product has proven to be valuable in situations where there is no indication regarding the identity of a bacterial organism. In bacteria, three genes are important which make up the rRNA functionality, that is 5S, 16S and 23S rRNA. The 16S rRNA gene has been most commonly employed for identification purposes because it is highly conserved and having a moderate copy number depending on the genus.

More recently, employment of the 16S-23S rRNA intergenic spacer region has become popular due to its high copy number and more importantly its high sequence variability. Primers are directed to highly conserved regions of the 16S and 23S rRNA genes and these may be either universal or specific, targeting a specific genus. Recently, 23S rRNA was used to identify bacterial species. The 23S rRNA locus shows more variation between species of medical importance than the 16S rRNA locus. Universal 23S rRNA primers were designed which allowed detection of bacterial agents from 158 positive blood cultures which were identified using a hybridization assay with specific oligonucleotide probes. It may be concluded that accuracy, range and discriminatory power of this assay could be continually extended by adding further oligonucleotides to the panel without significantly increasing complexity and cost.

Panbacterial qPCR amplification and sequencing of the gene encoding 16S ribosomal RNA using universal primers has become widely used in clinical laboratories for identification of bacteria at the genus and species levels, either after their isolation in culture or directly from clinical samples. This method helps in fast detection of wide spectrum of bacteria, especially in culture negative cases due to previous administration of antibiotics. This method is usually helpful when there is a single bacterial species in the clinical sample and the bacterial load is high enough to be detected. When there is multiple bacterial species in a clinical sample, a mix of 16S ribosomal DNA (rDNA) sequences will be obtained, and identification of individual species usually necessitates specific techniques (DNA cloning, high-throughput sequencing, ESI mass spectrometry etc.); however, these techniques are not available in many clinical laboratories. Notably, commercial software potentially allowing specific identification of mixed 16S rDNA sequences is available which could be used for testing most of the clinical samples except those contaminated with commensal flora. This could be applied for detection of bacteria in case of meningitis, endocarditis, arthritis, osteitis, and deep abscesses and suppurations. Panbacterial qPCR has also been used for quantification of the bacterial flora in sample including blood, intraocular samples or chronic wounds.

6.3 Quantitative endpoint PCR

PCR and reverse transcription PCR are used generally in a qualitative format to evaluate biological samples. However, a wide variety of applications, such as determining viral load, measuring responses to therapeutic agents and characterizing gene expression, are improved by quantitative determination of target abundance. Theoretically, this should be easy to achieve, given the exponential nature of PCR, because a linear relationship may exist between the number of amplification cycles and the logarithm of the number of molecules. In practice, however, the amplification efficiency is decreased because of contaminants (inhibitors), competitive reactions, substrate exhaustion, inactivation of the polymerase and target reannealing (Fig. 6.1). As the number of cycles increase, the amplification efficiency decreases, eventually resulting in a plateau effect.

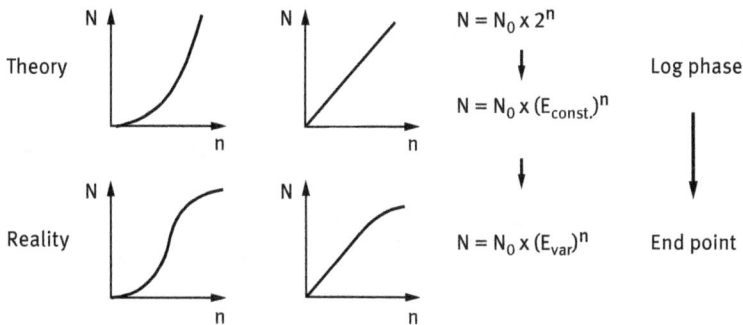

$$N = N_0 \times 2^n$$

$$N = N_0 \times (E_{const.})^n$$

$$N = N_0 \times (E_{var})^n$$

Theory / Reality / Log phase / End point

Fig. 6.1: Quantitative PCR: theory and reality. N, number of molecules amplified; n, number of cycles; N_0, original number of molecules; E, amplification efficiency.

Normally, quantitative PCR requires the measurements to be taken before the plateau phase so that the relationship between the number of cycles and molecules is relatively linear. This point must be determined empirically for different reactions because of the numerous factors that can affect the amplification efficiency. Because the measurement is taken prior to the reaction plateau, quantitative PCR uses fewer amplification cycles than basic PCR. This can cause problems in detecting the final product because there are fewer products to detect.

To monitor amplification efficiency, applications are designed to include an internal control in the PCR. One such approach includes a second primer pair that is specific for a "housekeeping" gene (i.e. a gene that has constant expression levels among the samples compared) in the reaction. Amplification of housekeeping genes, which serves as heterologous internal control, verifies that the target nucleic acid and reaction components were of acceptable quality but does not account for differences in amplification efficiencies due to differences in product size or primer annealing efficiency between the internal control and target being quantified.

A variation of quantitative PCR, known as competitive PCR, is a response to this limitation. A known amount of a control template is added to the reaction in competitive PCR and this template is amplified using the same primer pair as the experimental target molecule but yields a distinguishable product (e.g. different size, restriction digest pattern, etc.). The amount of this control, also known as homologous internal control, and test product are compared after amplification. While these approaches control for the quality of the target nucleic acid, buffer components, and primer annealing efficiencies, however, they have their own limitations.

Numerous fluorescent and solid-phase assays have been described to measure the amount of amplification product generated in each reaction, but they can fail to discriminate amplified DNA of interest from nonspecific amplification products. Some of these analyses rely on blotting techniques, which introduce another variable due to nucleic acid transfer efficiencies, while other assays have been developed to eliminate the need for gel electrophoresis yet provide the requisite specificity.

6.4 Real-time PCR (qPCR)

The qPCR, also called kinetic PCR, is a PCR-based laboratory technique, which is used simultaneously to amplify and quantify a DNA target. It helps both detection and quantification of a specific sequence in a DNA sample. It follows the general principle of PCR. With this method, the amplified DNA is quantified as it accumulates in the reaction in real-time after each amplification cycle. The qPCR can be combined with reverse transcription to quantify RNA or mRNA in cells or tissues.

Relative concentrations of DNA present during the exponential phase of the reaction are determined by plotting fluorescence against cycle number on a logarithmic scale (so an exponentially increasing quantity will give a straight line). A threshold for detection of fluorescence above background is determined. The cycle at which the fluorescence from a sample crosses the threshold is called the cycle threshold or crossing point. To quantify gene expression accurately, the measured amount of RNA from the gene of interest is divided by the amount of RNA from a housekeeping gene measured in the same sample to normalize for possible variation in the amount and quality of RNA between different samples. This normalization permits accurate comparison of expression of the gene of interest between different samples, if the expression of the reference (housekeeping) gene used in the normalization is very similar across all the samples. Choosing a reference gene fulfilling this criterion is therefore of high importance, and often challenging, because only very few genes show equal levels of expression across a range of different conditions or tissues.

6.5 Reporting results

Medical reports are specific formal documents provided by a laboratory to the referring physician or health-care professionals. The main target of reports is to provide a clear, concise, accurate, fully interpretative, and authoritative explanation to a clinical question as treatment of a patient mainly depends on the report obtained from the laboratory. Modern developments have shown an increased interest in utilizing measurements for the detection of trace levels of DNA. Such low-level detection methods are central to regulatory, public health, medical and quality control issues. There is little in the way of standardization of data handling from these methods, and the data generated need to be analyzed appropriately if the results are to be interpreted correctly. For some molecular tests, results need to be interpreted by qualified personnel.

There are two categories of results when using a qualitative assay: "detectable" (or "positive") or "not detectable" (or "<LOD"). It is of major importance to add always the LOD of the assay used (see Section 3.4.2.3). In some cases, further explanation is required for the interpretation of the result obtained and its clinical relevance. In contrast, there are three categories of results when using quantitative qPCR (Fig. 6.2). (1) If the value obtained exceeds the upper limit of quantification (ULQ), the value is reported as ">analytical measuring range". (2) If the value obtained is within the analytical measuring range, the value is reported quantitatively, along with the unit of measurement. It is of major importance to state always the analytical measuring range of the assay used in the report. (3) If the value obtained is below the lower limit of quantitation (LLQ), the result reported may be detectable but not accurately quantifiable ("positive <LLQ") or "not detectable". For instance, if plasma HIV-1 RNA is detected but the value is below the LLQ, residual HIV-1 viremia is present but not quantifiable.

The reproducibility of a molecular test or test system is another important parameter, which helps the physician to determine whether two consecutive results of a particular quantitative test are considerably different from each other. Information

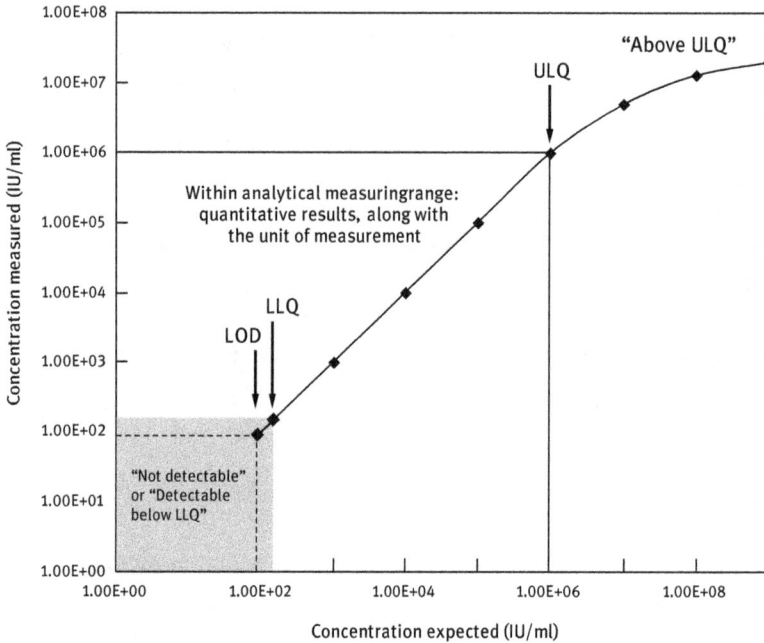

Fig. 6.2: Categories of results obtained with qPCR. In this example, the linearity ranges from 1.1E + 02 IU/ml to 1.0E + 06 IU/ml. Note that the LLQ may be equivalent to the LOD or it may be even at a higher concentration as shown here.

about reproducibility of the assay can be sent along with the report or it should be easily available to the physician. Furthermore, it is useful to report whether a detected microbe is interpreted as pathogen or normal flora. False-positives or false-negatives are the main disadvantages of most of the laboratory results. It would be important to address these limitations in the report depending on the nature of each assay and the significance of an incorrect result.

Some molecular reports may include assay sensitivity and specificity. The sensitivity report helps to identify how good the assay is in detecting a particular target DNA, and how good the assay is in detecting the disease associated with it. When the assay is not 100% sensitive for detecting a particular disease, it should be reported. The British Clinical Molecular Genetics Society and the Organisation for Economic Cooperation and Development have outlined Guidelines for executing and reporting molecular genetic test results. The American College of Medical Genetics has also devised guidelines for clinical genetics laboratories. Every laboratory must do studies to show how good each assay is performed in its facility. Information on assay performance characteristics should be made available on request but need not to be included in every patient report.

If the test result obtained from a specimen is different from that verified for this test, this has to be indicated clearly. Such a result must not be used for any therapeutical decision.

6.5.1 Genetic names

The American Society for Microbiology originally developed the rules, which are used for bacterial gene names in most clinical and research applications today. Usually, bacterial genes are named by three small letters defining the class of gene and a capital letter in the fourth position, which stands for the specific gene (e.g. *pgpA*, *pgpB*). Other organisms can have homologous genes, which have the same name, and so it is important to define the particular genus and species in the report. A collection of bacterial names with specific genomic reference sequences (RefSeqs) is found in the website www.ncbi.nlm.nih.gov/genomes/lproks.cgi. Each RefSeq gives the details of the name of genes its DNA sequence and amino acids. The nucleotide number and the amino acid substitution by the codon number indicate mutations. Normally names of genes are given in italics, but it is not always possible. In the list of viral RefSeqs at www.ncbi.nlm.nih.gov/genomes/static/vis.html, Epstein-Barr virus is called as human herpesvirus 4 (HHV 4) and human cytomegalovirus as human herpesvirus 5 (HHV 5). It is adequate to use the common medical name for each organism in clinical reports. The International AIDS Society at http://iasusa.org/resistance_mutations/index.html may be contacted for reports regarding HIV mutations and genotypes.

One should not give new names for genes or disease conditions, which already have one in the literature. The report should never confuse the doctor. However, when a name has not been defined, the informal names become more established and it would be difficult to make a change to the approved nomenclature system. During this period, it would be ideal to use the correct names followed by informal ones.

The US National Center of Biotechnology Information is trying to develop a reference sequence which could help in reporting clinical results and it named "RefSeqGene". Before it becomes available, it is advisable to give reports with genetic names the way in which clinicians are used to. Scientists are trying to make a base for the preparation of nomenclature of names for genes. Once named, the names and symbols for genes are not changed. For proper patient care, a structure is required for reporting genetic results in order.

6.5.2 Recommendations for reporting results of molecular tests

–   Follow reporting guidelines in checklists of the College of American Pathologists including descriptions of the patient, sample, testing laboratory, method and result.
–   Limit abbreviations to those described herein or to those defined on first use in a given report.
–   Interpret raw data sufficiently to facilitate clinical decision-making.
–   Discuss pertinent limitations of the test.
–   Keep the report concise and effective.
–   Report only validated assays and sample types.

- Do not report results for controls except if relevant to interpretation.
- Encourage clinical research so decisions are evidence-based.
- Use proper gene nomenclature and encourage its use system-wide.
- Encourage development of "RefSeqGene" to create a single reference sequence for each human gene (as well as for each human pathogen's genes).

6.5.3 Recommendations for the contents of the molecular test report

1. Laboratory/patient/sample identifiers
   (a) Name and address of reporting laboratory (optional: phone, fax, email, website)
   (b) Patient's name (first and last with middle initial or middle name)
   (c) ID No. and/or date of birth
   (d) Date of specimen collection (and time, if appropriate)
   (e) Date of receipt or accession in laboratory, with accession number
   (f) Specimen source (even if only one sample type is accepted) and how tissue was received (fresh is assumed unless designated as frozen, paraffin-embedded, etc.)
   (g) Ordering physician
2. Results
   (a) Results listed by test name; use standardized gene nomenclature and standard units of measure
   (b) Reference range; or normal versus abnormal
3. Interpretation
   (a) Analytical and clinical interpretation of results
   (b) Analytical interpretation involves synthesizing raw data to produce a reportable result
   (c) Clinical interpretation involves synthesizing analytical and clinical information to describe what the result means for the patient
4. Comments
   (a) Significance of the result in general or in relation to this patient
   (b) Correlation with prior test results
   (c) Recommendation of additional measures (e.g. further testing, genetic counseling)
   (d) Condition of specimen that may limit adequacy of testing (e.g. sample received thawed, partially degraded DNA)
   (e) Pertinent assay performance characteristics or interfering substances
   (f) Residual risk of disease (or carrier status) by Bayesian analysis
   (g) Control test results, if unusual or especially pertinent
   (h) Cite peer-reviewed medical literature or reliable websites on the assay and its clinical utility (e.g. educational materials on www.genetests.org)
   (i) Document intradepartmental consultation

(j) Document to whom preliminary results, verbal results, or critical values were reported and when

(k) Information specifically requested on the requisition (e.g. ethnicity)

(l) Reply to specific questions posed by the requesting clinician (e.g. rule out chronic lymphatic leukemia)

(m) Reason specimen rejected or not processed to completion

(n) Disposition of residual sample (e.g. sample repository)

(o) Chain of custody documentation, if needed

(p) If the report is an amended or addendum report, description of changes or updates

(q) Discrepancies between preliminary and final reports

(r) Name of testing laboratory, if transmitting or summarizing a referral laboratory's results

5. Procedure

(a) Type of procedure (e.g. PCR, reverse transcription PCR, qPCR, *in situ* hybridization, gene dosage array, RNA expression array, protein expression array, sequencing, protein truncation test)

(b) Defined target (i.e. name of analyte tested such as gene, locus, or genetic defect; use HUGO-approved gene nomenclature)

(c) Details of procedure, for example analyte-specific reagent or IVD/CE-labeled or FDA-approved kit including version and manufacturer, instrument type

(d) Disclaimer on laboratory-developed tests

(e) Signature and printed name of reporting physician, for any test having a physician interpretation

(f) Signature of lab director or designee when interpretation is performed (Reports may be signed electronically.)

(g) Date of report (and time, if appropriate)

6. Demographic information

(a) Accession number and specimen number from referring laboratory

(b) Genetic counselor, when appropriate

(c) Clinic/inpatient location; or name/address/phone of outside facility

(d) Indications for testing (reason for referral)

(e) Clinical history (clinical situation, ethnicity/race, pedigree diagram and/or family history, previous molecular/genetic studies and other relevant clinicopathologic findings)

6.6 Interpretation

Analytical interpretation and clinical interpretation are the two kinds of interpretation. Analytical interpretation involves examining the raw data and producing a reportable result. Clinical interpretation involves describing what the result means for the patient; either in general or based on specific knowledge of that patient's situation. Molecular tests are comparatively complex and are done by a variety of methods. Interpretation helps to confirm that the result is clinically significant and that the analytical limitations of the assay are revealed in relation to the findings. Especially doctors who often lack sufficient time need interpretations and resources for optimal judgment of the current literature with regard to use of laboratory tests. Laboratories can help in such cases by preparing reports that not only provide an analytical interpretation of the raw data but also provide guidance on the result for patient management. The principles of evidence-based laboratory medicine should be followed as summarized on the International Federation of Clinical Chemistry and Laboratory Medicine website (www.ifcc. org). The question should be answered in a concise and clear way, after taking into account any appropriate additional information supplied. Furthermore, it is recommended that laboratories highlight this conclusion (bold, underline, large font, text in a box etc.). In particular, it should be remembered that negative results (e.g. "no mutation detected") can easily be misinterpreted by non-specialists as "exclusions".

Only individuals who are proficient in the evaluation of the test analytically and clinically should interpret assays. Physicians believe that if interpretation is given along with the test result, it helps in both the differential diagnosis and the prevention of misdiagnosis. General comments included in a report should be concise and should have clinical significance. In addition, it is possible to include a source such as a website/publication in which the generic test information is found.

It is difficult to check the significance of a report if there are too many things included in it. It is advisable to give a summary of the test results at the top of the report. If possible, it is good to limit the report to a single page and it is advisable to take comments about the usefulness of the report from the physician. There can be a misunderstanding between the technician and the doctor in regards to with the words used in the report. It would be good if the doctors have an idea about the common terminologies used in reports. Normally this is a problem with junior doctors. This can be avoided if all reports are written properly keeping in mind that the reports are read not only by doctors but also by others such as nurses, researchers and even patients. It is logical to include references of published journals or the website of the testing laboratory. For many types of laboratory tests, both technical and clinical information may be available at websites.

Usually both external and internal controls are included with all clinical laboratory tests. External controls help in assuring the performance of the assay is as expected. Internal controls are used to determine the quality of the patient sample. Results of control samples are usually not reported. A clear statement regarding the significance of results specific to the clinical question should be given and it should belong to one of the following:

- Normal findings: Findings within the physiological variation which could be detected for the given individual specimen in view of the patient's ethnicity, age, sex, maturity, and other pertinent factors. It may be important to provide an estimate of the diagnostic sensitivity, the proportion of affected individuals likely to be detected. It could be useful to provide a key reference to support sensitivity estimates.

- Non-specific findings without clinical importance: There are findings other than physiological variation but not related to a common disease. These findings should be reported only if they appear to be relevant to the result.
- Incidental findings with possible clinical importance: These are findings pointing to a clinically significant question which is not related to the question which was asked. Decision about reporting these types of findings will depend on local policy and on how the patient has been counseled about this possibility. A clear policy on reporting incidental findings should be in place.
- Findings of uncertain significance: These are findings outside of the physiological variation but with possible importance to the clinical question asked. It is important that findings of uncertain significance are included in reports, as their significance may become clear at a later date. The possible significance of the finding should be explained by using literature, additional databases, prediction programs, or other means. In some cases, it may be important to provide an estimate of diagnostic specificity, which indicates the risk of a false-positive result.
- Disease-specific pathological findings: These findings are outside of the physiological variation and unquestionably associated with a specific clinically relevant disease. Pathogenicity of findings should be clearly supported from data in literature or databases. If any other information could be supplied or obtained from the referring clinician that might improve the accuracy of interpretation, it should be stated in the report.
- Interim reports: It may be useful to issue a report before all studies are complete. Interim reports should be clearly marked as such and should be worded to avoid misinterpretation of their status. Thus, phrases or summary statements appearing to give a definitive result should not be used. It should be clearly stated which analyses are still underway. The definitive report should clearly state which are the new results, and should include a general conclusion, taking account of all results.

6.7 Important issues when clinically interpreting molecular diagnostic results

For monitoring anti-HBV, anti-HCV, and antiretroviral therapies, determination of viral load in plasma or serum allows assessment of viral replication. When using different assays based on qPCR, quantities obtained have been more comparable recently, due to improved standardization with WHO standards. However, different quantitative molecular assays for detection of pathogens may produce significantly different results because of different extraction and amplification efficiencies, and by the use of proprietary standards for kit calibration. This reinforces the need for development of additional quantitative reference standards that could be used for calibration. Furthermore, it is strongly recommended to measure the viral load by the same molecular assay used on a patient.

Given the unique sensitivity of PCR, it may be noted that the molecular assay is detecting latent virus without any clinical significance. This has shown to be a problem especially when monitoring herpesviruses in immunosuppressed patients. Cut-off values for this have not yet been defined; a significant increase within a certain period is used instead for diagnosis.

Finally, it is worthy to note that even after successful therapy (pieces of) nucleic acids may circulate and be detected by a molecular assay. However, this does not necessarily mean unsuccessful therapy. Nevertheless, the clinical presentation must always be taken into consideration.

6.8 Take home messages

- Physician reports should provide a clear, concise, accurate and fully interpretable answer to a clinical question. When reporting results, the method(s) used should always be mentioned.
- There are two categories of results when using a qualitative assay: "detectable" (or "positive") or "not detectable" (or "<LOD"). It is of major importance to add always the LOD of the assay used.
- There are three categories of results when using quantitative qPCR: ">analytical measuring range", a quantitative value (along with the unit of measurement), and "positive <LLQ" or "not detectable". It is of major importance to state always the analytical measuring range of the assay used.
- Only individuals who are qualified for validation of test results should interpret assays.
- For monitoring viral load, it is strongly recommended to monitor a patient by the same assay.

6.9 Further reading

[1] Cobo, F. (2012) Application of Molecular Diagnostic Techniques for Viral Testing. Open Virol. J. 6:104–114.

[2] Espy, M.J., Uhl, J.R., Sloan, L.M., Buckwalter, S.P., Jones, M.F., Vetter, E.A., Yao, J.D.C., Wengenack, N.L., Rosenblatt, J.E., Cockerill, F.R. III, Smith, T.F. (2006) Real-time PCR in clinical microbiology: applications for routine laboratory testing. Clin. Microbiol. Rev. 19:165–256.

[3] Gulley, M.L., Braziel, R.M., Halling, K.C., Hsi, E.D., Kant, J.A., Nikiforova, M.N., Nowak, J.A., Ogino, S., Oliveira, A., Polesky, H.F., Silverman, L., Tubbs, R.R., Van Deerlin, V.M., Vance,G.H., Versalovic, J.; Molecular Pathology Resource Committee, College of American Pathologists (2007) Clinical laboratory reports in molecular pathology. Arch. Pathol. Lab. Med. 131:852–863.

[4] Kommedal, Ø., Simmon, K., Karaca, D., Langeland, N., Wiker, H.G. (2012) Dual priming oligonucleotides for broad-range amplification of the bacterial 16S rRNA gene directly from human clinical specimens. J. Clin. Microbiol. 50:1289–1294.

[5] Millar, B.C., Xu, J., Moore J.E. (2009) Molecular diagnostics of medically important bacterial infections. Curr. Iss. Mol. Biol. 9:21–40.

7 Human immunodeficiency virus

Jacques Izopet

Infection with the Human immunodeficiency virus (HIV) that causes acquired immunodeficiency syndrome (AIDS) was reported first in the United States in 1981. HIV-1 was isolated in 1983 by Françoise Barré-Sinoussi and Luc Montagnier (Nobel Prize in 2008), and HIV-2 was isolated in 1986. More than 25 million people have succumbed to the infection and 33 million are currently living with HIV. The HIV-1 virus has spread throughout the world and has become the most common cause of death in Africa. In contrast, HIV-2 infection is found primarily in West Africa and it is less prevalent worldwide than HIV-1. Successful treatment has made HIV/AIDS a manageable chronic disease in developed countries, but there is still no cure and protective vaccination remains elusive.

HIV-1 and HIV-2 belong to the *Lentivirus* genus in the family Retroviridae. They are enveloped RNA viruses with the following genes: *gag*, encoding capsid proteins; *pol*, encoding RT, protease (PR) and integrase (IN); and *env*, encoding the surface envelope glycoprotein (SU) and transmembrane envelope glycoprotein (TM) that allow the virus to bind to and enter target cells that bear CD4 (main receptor) and CCR5/CXCR4 chemokine receptors (coreceptors). HIV-1 and HIV-2 also have accessory genes (*tat, rev, nef, vif, vpr, vpu/vpx*) that determine their replicative and pathogenic characteristics.

HIV-1 viruses belong to one of four groups, the M (major), O (outlier), N (non-M, non-O), and P groups (Tab. 7.1). Group M includes subtypes also referred to as clades A to K, with B being the predominant subtype in Europe, North America and Australia, and C the most commonly transmitted virus. A number of recombinant viruses that form by exchanges in several regions of the virus between HIV-1 subtypes have been found additionally. HIV-2 viruses are classified into eight groups. HIV-1 and HIV-2 emerged because of transmission of simian immunodeficiency viruses (SIV) to humans. HIV-1 M and N are derived from SIVcpz that infects chimpanzees while HIV-1 O and P are related to SIVgor that infects wild-living gorillas, has been described recently. HIV-2 is derived from SIVsm that infects sooty mangabeys.

The virus strains can be classified as R5, X4, or R5X4 variants according to their use of one or both chemokine receptors CCR5 and CXCR4 to enter cells in addition to the CD4 molecule. The interaction of the SU glycoprotein with the CD4 receptor and one coreceptor results in rearrangement of the TM glycoprotein, which proceeds to fuse the membranes of the virion and the target cell. This permits the HIV nucleocapsid to enter the cell. RT results in the synthesis of HIV DNA from HIV RNA, which then becomes integrated into the genome of the human host cells, most of which are activated. These HIV-infected cells become persistent reservoirs for the virus and sources of transmission. The late steps of the viral cycle (transcription, translation, assembly, maturation, and exit of virions) depend on viral and cell factors. R5 viruses

Tab. 7.1: Global epidemiology of HIV variants.

Group	Subtype	Geographic region	Prevalence
HIV-1			
M	A	East Africa, West and Central Africa, Russia	High
M	B	America, West Europe, Australia, Southeast Asia	High
M	C	South and East Africa, India	High
M	D	East, Central and West Africa	High
M	E	Southeast Asia	High
M	F	West and Central Africa, Romania	Low
M	G	West and Central Africa	Rare
M	H	West and Central Africa	Rare
M	J	West and Central Africa	Rare
M	K	West and Central Africa	Rare
M	Recombinant forms	Worldwide	Increasing
N		West and Central Africa	Low
O		West and Central Africa	Rare
P		West and Central Africa	Low
HIV-2			
A–H		West Africa	Low

predominate in early HIV infection, while viruses that use CXCR4 (X4 and R5X4 viruses) are more cytopathic and emerge later in infection. Two major determinants can drive this virus evolution. One is the frequency of cells bearing both CD4 and CXCR4 and the density of CXCR4 on target CD4$^+$ cells at the sites of HIV replication. The other is the host immune response, including restriction factors. There is still some controversy about the role of chemokine receptor usage in determining the rate of HIV disease progression.

HIV infection occurs following exposure to infected secretions, most commonly genital fluids or blood, as well as through mother-to-child transmission. At the target mucosal site, the virus interacts with dendritic cells that produce costimulatory molecules and present HIV to activated CD4$^+$ T lymphocytes that have been primed for productive infection. The virus can be transported to local lymph nodes within hours of mucosal exposure and viremia can be detected 4–9 days later. Studies comparing the biological and genetic properties of viruses from a single individual suggest that those in distinct compartments like the blood, lymphoid tissue, gastrointestinal tract, genital fluids, or the central nervous system can evolve independently. The heterogeneity of the virus in both untreated and treated individuals is a key parameter for understanding the pathophysiology of HIV infection, designing diagnostic tools and defining successful therapeutic strategies.

7.1 Major symptoms

7.1.1 Untreated individuals

The natural progress of HIV infection is classified into three phases: acute infection, clinical latency and AIDS (Fig. 7.1).

7.1.1.1 Acute infection

Acute HIV infection presents as a non-specific viral syndrome in up to 90% of people. Symptoms develop 2–4 weeks after exposure. The clinical infection is characterized by fever, lymphoadenopathy, pharyngitis and occasionally exudates, macular or papular rash and malaise. Mucosal ulceration is seen in a minority of patients and can be an important clue to diagnosis. Ulceration can occur in the oropharynx, the esophagus, or on the external genitalia. Neurological symptoms are common, ranging from headaches or aseptic meningitis to more severe meningoencephalitis. The median duration of symptoms is 14 days. Viral replication usually explodes and the depletion of CD4$^+$ T lymphocytes in the peripheral blood is related to the loss of gastrointestinal tract CD4 T cells, probably because of direct viral infection.

7.1.1.2 Clinical latency

Most patients develop some degree of control of the HIV infection after the first few weeks by developing HIV-specific cytotoxic T lymphocytes and, possibly, neutralizing antibodies. The level of the new viremia set point determines the subsequent rate of progression of the HIV disease. CD4$^+$ T lymphocytes are depleted gradually;

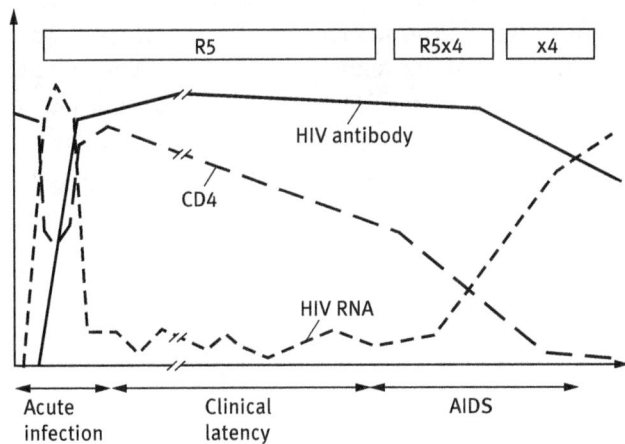

Fig. 7.1: Progression markers of HIV infection.

the average rate of decline is 70 cells/mm³/year. This decline is extremely variable, ranging from <200 cells/mm³ in as few as 2 years after acute infection to normal range indefinitely (elite controllers). Chronic systemic immune activation is a pathognomonic feature of progressive HIV infection and one of the strongest predictors of disease progression. Recent data indicate that the gut mucosal surface and the events that lead to its damage are crucial to the establishment of generalized immune activation.

7.1.1.3 AIDS

Untreated individuals whose CD4⁺ T cell counts drops to less than 200 cells/mm³ are subject to opportunistic infections and malignancies such as non-Hodgkin's and Hodgkin's lymphomas and Kaposi's sarcoma.

7.1.2 Treated individuals

Highly active antiretroviral therapy (HAART) has greatly reduced the morbidity and mortality of HIV-infected individuals. HAART can reduce the plasma viremia to levels that are undetectable by routine assays (<50 copies/ml) within 3 months in many patients. Importantly, HAART has been improved by the development of more potent and better tolerable therapies. However, the suppression of virus replication unveiled the existence of long-lived HIV reservoirs that involve very few cells, just 10^6–10^7 latently infected cells per individual, equivalent to 0.1–1 cell per million lymphocytes. Ongoing viremia can be detected using very sensitive tools. The majority of patients may have only 1–50 copies/ml. Although the origin of this viremia has not been fully characterized, the virions may engage CD4 and chemokine receptors and may activate pathways that could lead to long-term consequences, including cardiovascular dis-orders and malignant disease. The suboptimal penetration of many antiretrovirals into the central nervous system may also result in neuropathology. Three distinct mechanisms, not mutually exclusive, contribute to the persistence of HIV reservoirs. First, the intrinsic stability of latently infected CD4⁺ T cells is a major contributor. Second, the reservoir could be replenished continuously by slow viral replication that could lead to the *de novo* infection of target cells. Third, cell to cell transmission could also play a role.

7.2 Preanalytics

7.2.1 Specimen collection

In addition to immunoassays, nucleic acid testing (NAT) has become a crucial tool in the diagnosis of HIV infection in several circumstances. NAT is also the method of choice for monitoring HIV infection. HIV RNA or HIV DNA can be detected, quantified and characterized (resistance to antiretroviral drugs, tropism) from peripheral blood, or less frequently from other biological fluids such as genital fluids and cerebrospinal fluid (CSF).

The optimal blood collection parameters for testing the plasma HIV-1 viral load have been defined by many studies. There is now general agreement that ethylene-diamine-tetra-acetic acid (EDTA)-anticoagulated plasma is the most suitable and stable matrix for HIV-1 RNA quantification. Whole blood is processed usually within 12 h of specimen collection; however, it has been demonstrated that the HIV-1 RNA load remained stable for 72 h at 25°C. HIV RNA copy numbers are also not affected by long-term storage of EDTA plasma at –70°C. EDTA plasma is also used for testing drug resistance while HIV DNA is measured from EDTA whole blood specimens or in peripheral blood mononuclear cells (PBMC). Dried sample spots have been developed and validated as a method of collecting specimens in developing countries for monitoring HIV infection with reduced transportation costs. Dried blood spots and dried plasma spots are adequate for testing drug resistance and give results that are in good agreement with those obtained with frozen plasma. Several studies have also found viral loads by dried spot technology that were similar to those obtained by standard methods, but the small sample volume limits quantification to approximately 2000 copies/ml.

7.2.2 Clinical circumstances for using NAT to diagnose HIV infection

7.2.2.1 Primary HIV infection
Plasma HIV RNA can be detected earlier than the antibody or p24 antigen. An RNA test with a detection limit of 50 copies/ml detects HIV infection approximately 7 days before a p24 antigen test and 12 days before a third-generation anti-HIV antibody test.

7.2.2.2 Indeterminate serologic results
This situation occurs if the confirmatory (recombinant) immunoblot reaction does not meet the criteria for a clear-cut positive result. Testing for HIV-1 RNA can discriminate between early seroconversion and false reactivities due to antibodies to cellular components that are frequent in certain populations, such as multiparous women, multi-transfused patients, or patients with autoimmune diseases. People who have been given HIV trial vaccines may also produce an indeterminate confirmatory test. Testing for HIV-2 RNA with a non-commercial assay can identify a suspected HIV-2 infection.

7.2.2.3 Diagnosis of HIV infection in neonates
Nearly all the children of HIV-infected women have positive antibody tests due to the passage of maternal immunoglobulin G antibodies across the placenta. Uninfected

neonates can test positive for antibodies until they are 18 months old. It is important to establish the infection status of an HIV-exposed neonate because HAART has been shown to be effective in the very young. This can also help allay parental anxiety earlier. NAT is performed when a baby is less than 48 h old, aged 1–2 months, and aged 2–4 months after any prophylactic treatment has been stopped for at least 1 month. Studies have shown that detecting HIV-RNA in plasma and detecting HIV-DNA in PBMC or whole blood are all similarly effective.

7.2.2.4 Blood donor screening

HIV RNA assays have been developed to ensure the greatest possible safety of blood recipients. A multiplex format is often used that will also detect HCV and HBV. Testing for HIV RNA in the plasma of individual donors or mini-pools of donor plasma in addition to serologic testing has been shown to increase the number of infection cases identified significantly without impairing the performance of diagnostic testing.

7.2.2.5 Organ and tissue donors

Testing for plasma HIV RNA has replaced the HIV p24 antigen test in several countries, particularly for tissue donors when testing in an emergency context is not mandatory.

7.2.2.6 Artificial insemination

For artificial insemination, the semen of the HIV-1 positive male must be processed especially to remove HIV and must be tested for HIV nucleic acids before and after processing. The seminal plasma is tested for HIV RNA and the spermatozoa separated by the gradient density method followed by the swim-up method are tested for HIV RNA and HIV DNA.

7.2.3 Clinical circumstances for using NAT to monitor HIV infection

7.2.3.1 Viral load monitoring

The quantification of plasma HIV RNA is crucial for assessing potential disease progression and responses to therapy. Decisions regarding the initiation of antiretroviral therapy or changes to it are guided by monitoring the plasma viral load, the CD4$^+$ T-cell count and the patient's clinical condition. The goal of antiretroviral therapy is to reduce the plasma HIV RNA to a level that cannot be detected using sensitive assays. The viral load is monitored 1 month after the patient starts treatment, then at month 3, and every 3 months thereafter. The viremia should be reduced by at least 1 log$_{10}$ after 1 month and the plasma HIV RNA should have fallen below 50 copies/ml by 3–6 months. Changes in both the viral load and the CD4$^+$ T lymphocyte count are good predictors of clinical outcomes after the initiation of HAART.

The concentration of HIV DNA associated with PBMC has been shown to predict the evolution of untreated HIV-1 infections independently of the CD4$^+$ cell count and the plasma HIV-1 RNA concentration. The PBMC-associated HIV-1 DNA is a good estimate of the latent reservoir in patients with undetectable plasma HIV RNA. Therefore, the HIV-1 DNA in PBMC, together with the un-integrated and integrated forms of HIV-1 DNA, is the key marker for evaluating new therapeutic strategies for attacking the HIV reservoir to obtain a cure.

7.2.3.2 HIV drug resistance testing

The drugs used to treat HIV include nucleoside and nucleotide reverse transcriptase inhibitors (NRTI), non-nucleoside and nucleotide reverse transcriptase inhibitors (NNRTI), protease inhibitors (PI), integrase inhibitors (II), and entry inhibitors (EI).

The resistance of HIV to antiretroviral drugs can be transmitted via a new infection or a superinfection, or acquired when HIV replication is not completely suppressed, so allowing the HIV to evolve under the selective pressure of therapy. HIV drug resistance is therefore tested for three main categories of HIV infected patients: (1) patients presenting with a primary infection; (2) untreated patients who are chronically infected at the time of diagnosis or before treatment initiation; (3) treated patients with virologic failure, that is, patients with a detectable plasma HIV RNA despite HAART.

7.2.3.3 HIV tropism testing

There is a well-established relationship between coreceptor use and HIV disease progression. CCR5 antagonists are now available for treating patients with R5 viruses. HIV tropism must be assessed before treatment when the use of an anti-CCR5 is considered. The determination of HIV tropism in patients treated with an anti-CCR5 experiencing a virologic failure also can help to clarify how resistance to these drugs develops and guide clinical practice. There are two main pathways to resistance. The first involves expansion of pre-existing CXCR4-using viruses that are insensitive to anti-CCR5. The second involves continued use of CCR5 due to more efficient use of inhibitor-free CCR5 (competitive mechanism), or the use of the inhibitor-bound form of CCR5 (allosteric mechanism).

7.3 Analytics

7.3.1 Main technologies for NAT

The polymerase chain reaction (PCR), based on (c)DNA amplification, is the key technique used for the molecular diagnosis of HIV infection. The nucleic acids are extracted in a wide range of samples. The HIV RNA and HIV DNA are detected and quantified by the conventional method, in which the amplified DNA is detected after

Tab. 7.2: Frequently used commercially available assays for detection and quantification of HIV-1 RNA.

Manufacturer	Kit name	Amplification method	Target sequence	Range of linearity (copies/ml)
Gen Probe	Procleix Ultrio Assay	TMA	LTR + *pol*	Qualitative
Roche	COBAS AmpliScreen HIV-1, version 1.5	PCR	*Gag*	Qualitative
Abbott	Real Time HIV-1	qPCR	*pol* intregrase	4.0×10^1–1.0×10^7
bioMerieux	NucliSens Easy Q HIV-1, version 2.0	NASBA	*gag*	7.5×10^1–1.0×10^7
Qiagen	artus HIV-1 RT-PCR Kit	qPCR	LTR	1.2×10^2–1.0×10^8 [a]
Roche	COBAS AmpliPrep/ COBAS TaqMan HIV-1 Test, version 2.0	qPCR	LTR + *gag*	2.0×10^1–1.0×10^7
Siemens	Versant HIV-1 RNA 1.0 Assay	qPCR	*pol*	3.7×10^1–1.1×10^7

[a] IU/ml

amplification, or by (qPCR, in which the amplified DNA is detected during the PCR process. The tests for HIV drug resistance and tropism also rely on PCR.

Alternative technologies based on target amplification include transcription mediated amplification (TMA) and nucleic acid sequence based amplification (NASBA). TMA and NASBA are based on RNA amplification.

Most assays used in clinical practice are available commercially, but "homebrew" PCR assays may be useful for specific applications provided they have been verified carefully (see Chapter 3). Because many tests were initially designed for HIV-1 B subtypes, their performance for non-B HIV-1 subtypes must be carefully evaluated. Recent advances in instrumentation have led to improved automation of sample preparation and nucleic acid amplification, and hence to greater throughput. The introduction of internal quality controls in each run and the participation in an external quality control program are essential to ensure the continuing optimal performance of NAT.

Frequently used commercially available assays for detection and quantification of HIV-1 RNA are shown in Tab. 7.2.

7.3.2 HIV RNA assays

7.3.2.1 Qualitative tests
Commercial assays for HIV RNA in plasma have been developed for use in blood banks to screen blood donors. Similar assays could also be useful for a range of clinical

diagnostic purposes, including organ and tissue donors, artificial insemination, indeterminate serologic results and the diagnosis of HIV infection in neonates.

7.3.2.2 Quantitative tests

The first commercial assays were developed 20 years ago and newer versions with improved analytical performances and automation are now available. They offer a low limit of detection (<50 copies/ml), accurate quantification of non-B subtypes, wide linear range, and assays run in a close-tube format that minimizes the risk of contamination.

The conventional RT-PCR assay (COBAS Amplicor HIV-1 Monitor™, Roche) has been replaced by RT-qPCR assays. The COBAS AmpliPrep/COBAS TaqMan HIV-1™ (Roche) was the first platform to use this technology. Both RT and amplification primers and the probe are targeted to the sequence of a highly conserved region within the HIV-1 *gag* gene. The current version of the test (version 2.0), targeting highly conserved regions of both the HIV-1 *gag* gene and the HIV-1 LTR region, was developed for improved non-B HIV-1 quantification and group O detection. The platform includes an AmpliPrep instrument for automated sample preparation and either the COBAS TaqMan 48 analyzer or the COBAS TaqMan 96 analyzer for automated and high-throughput amplification and detection. The RealTime HIV-1 (Abbott) is run on two instruments, an automated sample preparation instrument and an integrated qPCR instrument. The primers and the probe are targeted to a sequence in a highly conserved region of the HIV-1 integrase of the *pol* gene and the assay is designed to detect HIV-1 group O. The Versant HIV-1 RNA 1.0 Assay, (Siemens) is performed on two instruments, a sample preparation module designed for automated nucleic acids extraction and an amplification/detection module designed for qPCR. The target sequence is a region of the *pol* gene. The HIV-1 RG RT-PCR kit (Qiagen) has recently become available. A low cost academic assay (Generic HIV, Biocentric) has also been developed.

The NucliSens Easy Q System™ (bioMérieux) is an automated system that combines a NASBA test and real-time molecular beacon detection. Molecular beacons have a stem-loop structure and contain a fluorophore and a quencher. The target is a region of *gag*.

The lack of a commercial assay for quantifying HIV-2 RNA in plasma has led to the development of "homebrew" assays, but their performance in terms of detection limit and accurate quantification vary according the different groups of HIV-2.

7.3.3 HIV DNA assays

The qualitative AMPLICOR® HIV-1 Test (Roche) is a manual *in vitro* test that utilizes PCR technology and nucleic acid hybridization to detect HIV-1 DNA in PBMC preparations. This test is optimized for Subtype B. It is for Research Use Only and is not available in all markets. The HIV DNA load can also be measured in whole blood or PBMC preparations with homebrew real time PCR assays.

7.3.4 HIV drug resistance assays

Frequently used commercially available assays for HIV drug resistance testing are shown in Tab. 7.3.

7.3.4.1 Genotyping methods

The main method used to analyze HIV drug resistance is genotype testing. Genotyping assays detect drug resistance mutations in relevant viral genes using a method based on line probe hybridization or DNA sequencing. The Inno-LiPA HIV-1 RT and PR tests (Siemens) use reverse hybridization with DNA probes on nitrocellulose strips that simultaneously detect wild-type codons and mutations in the HIV-1 RT and PR genes. The LiPA is designed to identify known primary mutations associated with high

Tab. 7.3: Frequently used commercially available genotyping and phenotyping methods for testing HIV-1 drug resistance.

(Kit) name	Manufacturer	Characteristics
Genotyping		
Inno-LiPA HIV-1	Siemens	Reverse hybridization RT (7 codons), PR (8 codons)
Trugene HIV-1	Siemens	Sequencing of RT (codons 40 to 247) and PR (codons 1 to 99)
ViroSeq HIV-1	Abbott	Sequencing of RT (codons 1 to 320) and PR (codons 1 to 99)
HIV sequencing	Homebrew	Sequencing RT, PR, IN and *env*
ANRS	http://hivfrenchresistance.org	Algorithm
Stanford	http://hivdb.stanford.edu	Algorithm
Rega	http://rega.kuleuven.be	Algorithm
Geno2Pheno	http://coreceptor.bioinf. mpi-sb.mpg.de/cgi-bin/ coreceptor.pl	Algorithm
Virtual phenotype	Monogram	
Virtual phenotype	Virco	
Phenotyping		
Phenosense	Monogram	
Antivirogram	Virco	

drug resistance, while sequencing can detect new mutations. The main genotyping technique is sequencing the genes associated with the specific antiretroviral drugs. It is also used for HIV-1 subtyping. Trugene HIV-1 genotyping (Siemens) uses RT-PCR and sequencing technology (CLIP) to sequence the entire PR gene and codons 40–247 of the RT gene. The ViroSeq HIV-1 genotyping system (Abbott) uses RT-PCR and dye terminator sequencing to determine the sequence of the entire PR gene and the first 320 codons of the RT gene. In addition, "homebrew" assays have been developed on different DNA sequencing platforms with capillary-based, semi-automated implementations of the standard dideoxy-nucleotide-terminator biochemistry. This requires more hands-on time but is very flexible and the cost per test is lower. Standardized protocols with PCR and sequencing primers for RT, PR, IN and Envelope, including reaction conditions are available.

The interpretation of HIV-1 drug resistance data relies on algorithms that analyze the correlation between drug resistance mutations and either susceptibility testing by phenotyping, or the virological outcome of patients who fail to respond to antiretroviral therapy. The rules are usually presented as tables listing mutations conferring resistance. These are updated regularly by a consensus panel using information available in the peer-reviewed literature, from clinical drug trials, and from research studies specifically designed to optimize existing algorithms. There are also algorithms derived from a database of genotype-phenotype methods that give a virtual phenotype. This is an estimation of the phenotype obtained by averaging viruses with similar genotypes from a large database of historical genotype-phenotype results.

Genotype testing is performed usually on plasma samples with viral load of >1000 copies/ml. The clinical relevance of genotype testing in patients with undetectable plasma HIV RNA using cell HIV DNA has not been fully evaluated.

7.3.4.2 Phenotyping methods

Phenotype testing is based on the Recombinant Virus Assay (RVA) developed by reference laboratories. The virus RT and PR genes obtained from a patient's blood sample are amplified by PCR and a recombinant virus is constructed using a cloned provirus from which the corresponding region of the genome has been removed. Recombinant–non competent virus is used in the PhenoSense HIV™ assay (Monogram) and recombinant-competent virus is used in the Antivirogram HIV™ assay (Virco) to infect the target cells used for testing the virus susceptibility to the various drugs. The resulting drug susceptibility curve is used to calculate the concentration of drug required to inhibit viral replication by 50% (IC 50) or 90% (IC 90). Phenotypic assay results are interpreted in terms of the specific resistance (i.e. fold increase in IC50 and IC90 with three categories 2–4, 4–10, and >10) that is associated with drug failure. Clinically significant fold increase cut-offs have been defined for each drug.

Phenotyping is technically complex, the test requires a long turnaround time and the cost is higher (three times approximately) than genotyping. In addition, the accuracy of clinical cut-off values for predicting a response remains somewhat doubtful. Lastly, there are few data supporting the benefit of phenotype testing in terms of clinical utility, except for a new drug when the pattern of mutations conferring resistance to this drug is still unknown.

7.3.5 HIV tropism assays

Frequently used commercially available molecular assays and interpretation algorithms for HIV tropism testing are shown in Tab. 7.4.

7.3.5.1 Genotyping methods
The major genotype determinants of HIV-1 coreceptor usage lie in the third variable loop (V3) of the SU envelope glycoprotein. Several studies have demonstrated that sequencing of the *env* V3 region DNA can discriminate between R5-using viruses and CXCR4-using viruses. This latter category includes pure X4 viruses, R5X4 viruses (dual, D), and mixtures of X4, R5 and R5X4 (mixtures, M).

Algorithms based on large sets of genotype-phenotype correlations, such as Geno2Pheno and Web PSSM, have been developed. Recent studies indicate that genotype rules based on the amino-acid residues at positions 11 and 25 and the overall net charge of V3 perform better for predicting the tropism for non-B HIV-1 subtypes including subtypes C, D, CRF01_AE, and CRF02_AG (see http://hivfrenchresistance.org).

Tab. 7.4: Frequently used commercially available genotyping and phenotyping methods for testing HIV-1 tropism.

(Kit) name	Manufacturer	Characteristics
Genotyping		
ANRS	http://hivfrenchresistance.org	Algorithm
Geno2Pheno	http://coreceptor.bioinf. mpi-sb.mpg.de/cgi-bin/ coreceptor.pl	Algorithm
PSSM	http://ubik.microbiol. washington.edu/ computing/pssm	Algorithm
Phenotyping		
Trofile™ES	Monogram	Sensitivity for CXCR4-using variants 0.3%

7.3.5.2 Phenotyping methods

Phenotype testing is based on the RVA developed by reference laboratories. These assays use RT-PCR to amplify the *env* gene from a plasma sample that is used then to produce recombinant viruses expressing the challenge *env* gene. Indicator cell lines bearing CD4 and either CCR5 or CXCR4 coreceptors are then infected. Clinical trials of CCR5 antagonists have relied on determining the phenotype of HIV-1 tropism with a commercial assay (Trofile™, Monogram). This assay detects minor CXCR4-using variants when they account for 10% in the virus population. A new sensitive version, the enhanced Trofile™ assay, detects 0.3% CXCR4-using variant. Cost limitations and long turnaround times have led to the development of "homebrew" RVAs such as the Toulouse Tropism Test that have been validated for both plasma samples and cells. Several studies have indicated good agreement between the results obtained with the various phenotype tests.

7.3.6 Assays for minority HIV variants

An HIV infection is caused in fact by a swarm of genetically diverse strains of virus. Direct sequencing using the standard dideoxy-nucleotide-terminator biochemistry used in conventional genotype assays can only detect nucleotide polymorphisms that account for ≥20% of a clinical sample. However, recent data suggest that resistant viruses that make up as little as 1% of the virus population can grow rapidly under the selective pressure exerted by drugs and lead to virologic failure. Similarly, CXCR4-using viruses can be detected at low frequency prior to CCR5 antagonist treatment. However, the frequency above which the response to therapy is significantly impaired remains to be defined. Three main techniques can be used to detect and quantify low-frequency HIV variants. First, allele-specific qPCR can detect variant viruses that represent as little as 0.01% of the population. The limitation is that multiple PCRs must be performed to analyze relevant HIV mutations. Second, DNA sequencing of multiple molecular clones after cloning the sequence of interest or after limiting dilution can discriminate variants present at 1–3% of the population, but it is time consuming and costly. Third, next-generation sequencing techniques using massively parallel sequencing platforms including those manufactured by Illumina, life technologies (Ion Torrent), and Roche (454 Sequencing) are rapidly developing. Next-generation sequencing with the 454 platform allows to read lengths of >400 bp providing a fantastic opportunity to analyze HIV diversity and to track the dynamic evolution of a mutant in response to selection pressure. The most promising clinical applications of next-generation sequencing are the detection and quantification of minor resistant variants to NNRTI and CXCR4-using viruses.

7.4 Postanalytics

7.4.1 Molecular diagnosis of HIV infection

Although HIV infection is usually diagnosed by testing for HIV-specific antibodies, NAT is crucial for diagnosis under certain circumstances. Results for neonates of seropositive mothers must be interpreted taking into consideration any antiretroviral therapy that has been given to the mother during labor by intravenous infusion and orally to the neonate for 4–6 weeks to prevent mother-to-child transmission. Therefore, the definitive un-infected status of the child cannot be established before therapy has been stopped. The assay used to diagnose infection of a baby born to a mother who is infected with HIV-2 or HIV-1 group O must be appropriate. In patients with indeterminate serologic results, interpretation of a negative NAT must also take into consideration the diversity of HIV variants, including those that could not have been detected. The same is true for patients suspected of having a primary infection. In all these situations, the test used must have excellent analytical performance in terms of sensitivity and specificity.

Identifying HIV-2 and group O HIV-1 infections is very important because these infections are intrinsically resistant to NNRTIs.

7.4.2 Monitoring HIV infection

The smallest change in viral load that is considered to be statistically significant (2 standard deviations) is a three-fold change, equaling 0.5 log copies/ml. Smaller changes are attributed to technical or biological variations. There may also be differences between assays or generations of assays. This may have an impact on the detection of samples close to the limit of detection or the lower limit of quantitation. Therefore, the single patient should always be monitored with the same assay to limit this restriction. The providers of assay systems need to improve the standardization of assay calibrations. Despite the good performance of NAT and the significant improvements achieved with newer versions, the genetic diversity of HIV is still a critical issue for accurate quantification. Plasma HIV-RNA concentration of rare variants can be underestimated and this must be suspected when there is any disagreement with the clinical condition or the $CD4^+$ T cell count.

The key goal of antiretroviral therapy is the reduction of the viral load under 50 copies/ml. The interpretation of drug resistance tests requires the collaboration of a multidisciplinary team of clinicians, virologists and pharmacologists. This is essential for designing a potent optimal regimen. The patient's adherence to treatment must be checked if the change in viral load is <1 log after 1 month. Similarly, detection of plasma HIV RNA in a patient after 3–6 months on HAART that has been selected and optimized by drug resistance testing before initiation may be due to suboptimal adherence or pharmacokinetic issues such as malabsorption or drug interactions.

Several studies also suggest that minority variants of drug-resistant viruses can be involved in early virological failure. This is a particular problem with low-level NNRTI-resistant variants. Very sensitive assays indicate that drug resistance occurs in 10–30% of patients with a primary infection; this is twice the figure obtained with standard population sequencing assays. Drug resistance mutations are often lost in the absence of selection pressure because they negatively affect virus replicative fitness. However, they can persist in reservoirs and be present at very low concentrations in naïve-patients. Nevertheless, further clinical evidence is needed to confirm the association between minority resistance mutations and HIV treatment failure. Very sensitive assays for routine resistance testing could have clinical utility for detecting low-level NNRTI-resistant variants. Recent data also indicate that HIV tropism testing with ultrasensitive assays (recombinant virus assays and next-generation sequencing) is clinically relevant.

The risk of failure is very low for a patient who adheres to the treatment regime and has a plasma HIV RNA level of less than 50 copies/ml. Resistant HIV variants cannot emerge and the large majority of patients have a durable response to therapy. Studies show that blips, or small low fluctuations in the viral load over the threshold of detection, usually have no clinical consequence. It is worth to note that very sensitive assays (detection limit <3 copies/ml) can detect low levels of virus in the plasma of patients whose virus has apparently been successfully suppressed for many years. Both residual ongoing virus replication in certain anatomical compartments and virus released from virus reservoir cells may contribute to the persistence of low viremia in patients on suppressive HAART. Although the virus replication seems to be slow enough to prevent the emergence of drug resistance, persistent low viremia could be responsible for the continued immune activation and the pathogenesis of HIV. The remarkable stability of the latent reservoir appears to be maintained by the intrinsic stability of latently infected resting CD4$^+$ T cells, renewal due to residual ongoing replication or cell to cell transmission, homeostatic proliferation, and the activation of latently infected cells. Strategies of treatment intensification in combination with activating agents to purge latent reservoirs are being evaluated currently. Therefore, the results of quantitative HIV DNA assays in patients with undetectable viremia are critical for clinical research.

7.5 Take-home messages

- HIV-1 viruses belong to one of four groups (M, N, O, P).
- The virus strains can be classified as R5, X4, or R5X4 variants according to their use of one or both chemokine receptors CCR5 and CXCR4 to enter cells.

- Serologic tests are used for screening, while NAT is crucial at primary infection stage, for indeterminate serologic results, and for diagnosis of HIV infection in neonates.
- NAT is also crucial for viral load monitoring, HIV drug resistance testing, and tropism testing.
- Main technologies are qPCR and HIV *pol-* or *env*-sequencing.
- Ultrasensitive assays (recombinant virus assays and next-generation sequencing) are key techniques for HIV tropism testing.

7.6 Further reading

[1] UNAIDS (2009) Report on the global AIDS epidemic: executive summary. http://www.unaids.org.
[2] Levy, J.A. (2009) HIV pathogenesis: 25 years of progress and persistent challenges. AIDS 23:147–160.
[3] Douek, D.C., Roederer, M., Koup, R.A. (2009) Emerging concepts in the immunopathogenesis of AIDS. Annu. Rev. Med. 60:471–484.
[4] Geeraert, L., Kraus, G., Pomerantz, R.J. (2008) Hide-and-seek: the challenge of viral persistence in HIV-1 infection. Annu. Rev. Med. 59:487–501.
[5] Richman, D.D., Margolis, D.M., Delaney, M., Greene, W.C., Hazuda, D., Pomerantz, R.J. (2009) The challenge of finding a cure for HIV infection. Science 323:1304–1307.
[6] Bushman, F.D., Hoffmann, C., Ronen, K., Malani, N., Minkah, N., Rose, H.M., Tebas, P., Wang, G.P. (2008) Massively parallel pyrosequencing in HIV research. AIDS 22:1411–1415.
[7] Wainberg, M.A., Zaharatos, G.J., Brenner, B.G. (2011) Development of antiretroviral drug resistance. N. Engl. J Med. 365:637–646.
[8] Saliou A., Delobel, P., Dubois, M., Nicot, F., Raymond, S., Calvez, V., Masquelier, B., Izopet, J. (2011) Concordance between two phenotypic assays and ultradeep pyrosequencing for determining HIV-1 tropism. Antimicrob. Agents Chemother. 55:2831–2836.
[9] Li J.Z., Paredes, R., Ribaudo, H.J., Svarovskaia, E.S., Metzner, K.J., Kozal, M.J., Hullsiek, K.H., Balduin, M., Jakobsen, M.R., Geretti, A.M., Thiebaut, R., Ostergaard, L., Masquelier, B., Johnson, J.A., Miller, M.D., Kuritzkes, D.R. (2011) Low-frequency HIV-1 drug resistance mutations and risk of NNRTI-based antiretroviral treatment failure: a systematic review and pooled analysis. JAMA 305:1327–1335.
[10] Loman N.J., Misra, R.V., Dallman, T.J., Constantinidou, C., Gharbia, S.E., Wain, J., Pallen, M.J. (2012) Performance comparison of benchtop high-throughput sequencing platforms. Nat. Biotechnol. 30:434–439.

8 Hepatitis viruses

Dieter Hoffmann, Thomas Michler and Ulrike Protzer

The term hepatitis describes an inflammatory disease of the liver. The most frequent origin worldwide is viral infection, but alcohol, drugs, other toxins, metabolic disorders, and autoimmune reactions may also cause hepatitis. Acute and chronic courses of viral hepatitis are observed. The clinical symptoms range from asymptomatic forms to fulminant liver failure.

Viral hepatitis is mostly caused by the hepatitis viruses A, B, C, D and E, but may also be related to herpesviruses (CMV, EBV and VZV; rarely HSV and HHV6) or adenovirus infection. Besides hepatitis A to E viruses (HAV to HEV), EBV causes hepatitis upon primary infection. In tropical areas, yellow fever virus as well as the hemorrhagic fever viruses (Marburg, Ebola, Lassa, Junin, Machupo, Kyasanur Forest) may cause liver inflammation.

Worldwide more than 350 million people are chronically infected with hepatitis B virus (HBV) and approximately 130 million with hepatitis C virus (HCV). Infections have similar clinical presentation; definitive diagnosis is dependent on laboratory testing. Besides serological studies, molecular assays are of major importance to confirm and stage the causative virus.

The qPCR, which has revolutionized nucleic acid detection by its speed, sensitivity, and reproducibility, is currently applied to the detection and quantitation of hepatitis viruses. Quantitative PCR plays a key role in laboratory diagnosis of non-A to -E hepatitis, in determining the risk of transmission, and in monitoring the response to antiviral treatment of hepatitis B, C and D.

HBV and HCV genotyping results provide important information to predict the resistance or response to antiviral therapy.

8.1 Major symptoms

Table 8.1 shows areas of infection and major symptoms caused by hepatitis viruses.

8.2 Preanalytics

HAV and HEV can be detected in stool specimens but also in serum and plasma. For molecular detection and quantitation of HBV, HDV, and HCV, serum or plasma specimens are used. Liver biopsy may be indicated to assess the degree of liver damage and inflammation in viral hepatitis and sometimes to rule out other differential diagnoses especially after organ transplantation but may also help to asses viral persistence, in

Tab. 8.1: Virus related areas of infection and symptoms.

Virus	Areas of infection	Symptoms
Adenovirus	Eye, gut, liver	Conjunctivitis, diarrhea, hepatitis
CMV	Liver, lung, retina	Swollen lymph nodes, encephalitis, hepatitis, retinitis, pneumonia
EBV	Tonsils, liver	Swollen lymph nodes, tonsillitis, fever, hepato-/splenomegaly, hepatitis, mononucleosis
HAV	Liver	Hepatitis, nausea, vomiting, diarrhea, fever, fatigue, abdominal pain
HBV	Liver	Hepatitis, appetite loss, fatigue, itching, nausea, abdominal pain;[a] liver fibrosis/cirrhosis,[b] HCC[b]
HCV	Liver	Hepatitis, fatigue, nausea, vomiting;[c] liver fibrosis/cirrhosis,[b] HCC[b]
HDV	Liver	Similar to hepatitis B, but often more severe
HEV	Liver	Hepatitis, diarrhea Severe symptoms during pregnancy possible
HHV-6	Liver, skin, CNS	*Roseola infantum*, fever, lymphadenopathy, hepatitis, encephalitis
HSV	Skin, CNS, liver	Small fluid-filled blisters, encephalitis, hepatitis
Lassa virus, Marburg virus, Ebola virus and other hemorrhagic fever viruses	Liver, kidney, spleen	Fever, myalgia, nausea, vasculitis, sore throat, hemorrhage, hepatitis
VZV	Skin, lung, CNS, liver	Exanthema, blisters, fever, pneumonia, meningoencephalitis, hepatitis
Yellow fever virus	Liver	Fever, headache, nausea, hemorrhage, hepatitis

[a] 50% without symptoms.
[b] Late complication of infection.
[c] 80% without or with only mild symptoms.

particular of HBV. However, specimens may also be collected from different body sites depending on the symptoms and the differential diagnosis (Tab. 8.2).

Freshly drawn whole blood specimens may be stored at 15–30°C for a few hours or at 2–8°C for up to 24 h. Plasma should be collected in EDTA or ACDA tubes. Heparin anticoagulated plasma is not suitable because heparin may inhibit PCR reactions. After centrifugation, separated plasma or serum may be kept at 15–30°C for up to 24 h or at 2–8°C for up to 1 week. For a longer period, they should be stored at –70°C or lower. Multiple freeze-thaw cycles should be avoided.

Tab. 8.2: Virus related specimens useful for detection of nucleic acids.

Virus	Specimens
Adenovirus	Plasma; serum, whole blood, tissue biopsy
CMV	Plasma, serum, whole blood, (urine), liver biopsy[a]
EBV	Plasma, serum, whole blood, liver biopsy[a]
Hepatitis viruses A–E	Plasma, serum, whole blood, stool (HAV, HEV), liver biopsy[a] (HBV, HDV, HCV)
HHV-6	Plasma, serum, whole blood, liver biopsy[a]
HSV	Plasma, serum, whole blood, liver biopsy[a]
VZV	Plasma, serum, whole blood, liver biopsy[a]
Lassa, Marburg, Ebola virus etc.	Plasma, serum (up to 2 weeks after onset), whole blood, (urine)
Yellow fever virus	Plasma, serum, whole blood

[a] Requires careful indication.

Stool specimens should be prepared freshly (within 1 h after collection) or stored at –70°C.

Liver biopsies should immediately be lysed for nucleic acid extraction or snap frozen and stored at –80°C. Fixed liver tissue can be used for molecular analysis, but at reduced yield. Paraffin-embedded tissue can also be used but paraffin needs to be removed before tissue lysis.

8.3 Analytics

Nucleic acid testing is of special importance in viral hepatitis to confirm diagnosis, to assess infection activity and the risk of transmission, to define the virus genotype and, to monitor and guide antiviral therapy.

Sensitive nucleic acid tests depend on efficient methods for RNA and DNA extraction starting from patient material. Especially for HBV detection, efficient lysis of viral particles is essential to release nucleic acids. Homologous or heterologous internal controls must be added before extraction to monitor both nucleic acid extraction and PCR.

Nucleic acid testing for detection of viruses causing hepatitis may be based on different methods (Tab. 8.3). Today, the most important is qPCR. Amplification is confirmed and quantified by hybridization with specific fluorescent probes (see also Chapter 5). However, if quantitation is not needed, conventional PCR may be run.

In addition to qPCR, alternative techniques are still being used but their importance is declining. Transcription-mediated amplification (TMA) amplifies RNA by *in vitro* transcription of a DNA template without thermal cycling. Besides reverse

Tab. 8.3: Nucleic acid testing methods recommended for detection of viruses causing hepatitis.

Virus	Nucleic acid	Recommended method
Adenoviruses	DNA	qPCR
CMV	DNA	qPCR
EBV	DNA	qPCR
HAV	RNA	Conventional RT-PCR or qRT-PCR
HBV	DNA	qPCR (ultrasensitive)
HCV	RNA	qRT-PCR (ultrasensitive)
HDV	RNA	qRT-PCR
HEV	RNA	Conventional RT-PCR or qRT-PCR
HHV6	DNA	qPCR
HSV	DNA	qPCR
VZV	DNA	qPCR

transcriptase, it involves primers adding a T7 recognition sequence and amplification by T7-RNA polymerase. Single-stranded, fluorescently labeled probes are then hybridized to the RNA amplification products allowing detection, whereas non-hybridized probes are chemically degraded.

For hepatitis viruses A to E, PCR allows early laboratory diagnosis of a viral hepatitis, even before seroconversion. Furthermore, it defines active infection and indicates infectivity by contact to stool in hepatitis A and E, by blood transfusion, stem cell or solid organ transplantation, by needle sharing for all hepatitis viruses or by sexual contacts or during birth from mother to child for HBV.

For herpes- and adenoviruses, qPCR is used to determine reactivation of the respective virus in immunosuppressed patients especially in transplantation settings (see also Chapter 9). For primary infection with EBV or with one of the exotic viruses able to induce liver inflammation, nucleic acid detection has lower priority.

Hepatitis B and C can be treated, sensitive quantitative viral load assays are thus required to monitor response to therapy and guide specific therapies. The viral genotype helps to predict treatment outcome and should therefore be determined before therapy is initiated. Emergence of viral resistance mutants during therapy is an important issue in antiviral therapy of hepatitis B, but may also become increasingly important for hepatitis C with the novel directly acting antivirals gaining clinical approval.

8.3.1 Adenoviruses

Adenoviruses cause hepatitis mostly under immunosuppression. Because of the large diversity of adenovirus serotypes and the high seropositivity rate in the population to date, the diagnosis is usually based on laboratory-developed PCR assays. PCR primers commonly target the hexon or the fiber gene. Careful selection of primers is required to guarantee detection of the more than 50 serotypes identified. Subtyping is not routinely performed even though subtype-specific clinical implications have been reported.

8.3.2 HAV

Molecular diagnosis of HAV is done in stool, serum, or plasma by conventional RT-PCR or RT-qPCR. Quantitation is not needed because neither chronic infection nor antiviral treatment exists. Molecular assays target conserved regions, such as the 5′-nontranslated (NTR), the polymerase or the protease region. Detection of HAV RNA in stool indicates infectivity of the patient. Several laboratory-developed assays have been published and commercially available kits have been brought on the market.

8.3.3 HBV

Even though primary diagnosis of HBV infection can usually be obtained by serological testing, sensitive PCR is used to diagnose very recent infection before hepatitis B surface antigen (HBsAg) is detected, infectivity during seroconversion from HBsAg to anti-HBs, and infections with HBsAg-virus mutants that escape serological detection. Furthermore, sensitive PCR helps to assess the risk of transmission in occult hepatitis B (e.g. isolated seropositivity for anti-HBc). Plasma and serum are the preferred sample types but liver or other biopsies can also be tested.

In persisting HBV infection (for more than 6 months), quantitation of the viral load is required to assess prognosis because the risk of developing hepatocellular carcinoma strongly correlates with HBV DNA titers. Therefore, quantitative assays are crucial for monitoring the infection and determining the indication for antiviral therapy. Available therapies include pegylated interferon-α and nucleos(t)ide analogues, which have to be applied long-term, maybe even lifelong. Therapy monitoring requires highly sensitive quantitative assays usually based on qPCR and amplifying part of the precore/core region or the S gene (detection limit ≤25 IU/ml). Quantitative results should be expressed as "IU/ml"; however, several assays still report results in "copies/ml". With slight variations from test to test, 1 IU/ml corresponds to 5 copies/ml. Current commercially available assays frequently used for the quantitation of HBV DNA are compared in Tab. 8.4.

HBV is classified into eight genotypes, A to H. Genotypes usually are determined by sequence analysis of the S gene or by using line probe assays. Specific PCR assays may also be applied.

Tab. 8.4: Comparison of currently frequently used commercially available assays for the quantitation of HBV DNA.

Characteristics	Manufacturer and details		
	Abbott	**Qiagen GmbH**	**Roche molecular diagnostics**
Kit name	RealTime HBV	artus HBV RG PCR Kit	COBAS AmpliPrep/COBAS TaqMan HBV Test[a]
Target sequence	HBs gene	Core region	Precore/core region
Amplification method	qPCR	qPCR	qPCR
Detection method	Fluorescence	Fluorescence	Fluorescence
Internal control	Heterologous	Heterologous	Homologous
Standards	Four EQS	Five EQS	One IQS
Analytical measuring range (IU/ml)	$1.0 \times 10^1 - 1.0 \times 10^9$	$3.2 \times 10^1 - 2.0 \times 10^7$	$2.0 \times 10^1 - 1.7 \times 10^8$

EQS: external quantitation standards; IQS: internal quantitation standard.

[a] The COBAS TaqMan HBV Test may also be used together with the High Pure System.

Long-term therapy with nucleos(t)ide analogs harbors the risk to select for drug resistance mutations within the RT region of the HBV polymerase gene. If virus titers rise during antiviral treatment, resistance testing should be initiated. Because phenotypic assays are expensive, time consuming and can only be performed in a few highly specialized laboratories, genotypic resistance testing has become state of the art. Known resistant mutants can be detected by line probe assays. Alternatively, sequencing of the RT region allows detection of known as well as novel mutations associated with drug resistance. Since cross-resistance to other antivirals is frequently observed and mutant viruses may be transmitted, genotypic resistance testing may also be indicated before initiating or adapting antiviral treatment. Since HBV cure becomes increasingly important as a clinical goal, testing of HBV cccDNA, the persistence form of HBV in hepatocytes, becomes interesting but is so far only possible with experimental assays.

8.3.4 HCV

The HCV was among the first viruses for which quantitative nucleic acid assays were established. Monitoring HCV RNA load by NAT is state of the art for several years. Plasma and serum are the preferred sample types but liver biopsies may also be tested. At present, RT-qPCR is by far the most commonly used technology and usually amplifies parts of the 5′ NTR.

Current therapies are based on pegylated interferon-α and ribavirin, which may be combined with directly acting antivirals. Currently, the protease inhibitors boceprevir and telaprevir are licensed for GT1. In 2014, two further protease inhibitors (simeprevir and faldaprevir) and the first polymerase inhibitor (sofosbuvir) are expected to be licensed by FDA and ESA. In contrast to HBV, HCV can be eradicated by antiviral therapy. The therapeutical proceeding depends on the response to therapy assessed by the decrease of virus load. Ultrasensitive molecular assays (with a detection limit of ≤15 IU/ml) are critical for response-guided therapy. To ensure ultrasensitive detection, commercially available assays based on qPCR are employed (Tab. 8.5).

HCV is currently classified into genotypes 1 to 6, which can be further differentiated into numerous subtypes. Prognosis and duration of antiviral therapy depends on the viral genotype; thus, genotyping is recommended before therapy is initiated. The 5′-NTR is the preferred part of the HCV genome for genotyping. It is sufficiently diverse between subtypes for differentiation and is highly conserved within subtypes and genotypes limiting "background noise". Available methods include sequencing of the 5′-NTR, line probe assays, and specific qPCRs (Tab. 8.6). Drug resistance testing is currently not used routinely but may become necessary when directly acting antivirals (DAAs such as protease and polymerase inhibitors) fail to eliminate the virus. Sequencing of the NS3 or/and NS5 region is necessary for genotypic drug resistance testing.

8.3.5 HDV

HDV can only infect HBV-positive individuals. In patients with positive anti-HDV antibody testing, detection of HDV RNA by RT-qPCR in serum or plasma proves active infection. HDV RNA quantitation may be useful in chronic HBV/HDV co-infections,

Tab. 8.5: Comparison of currently frequently used commercially available assays for the quantitation of HCV RNA.

Characteristics	Manufacturer and details		
	Abbott	QIAGEN	Roche Molecular Diagnostics
Kit name	RealTime HCV	artus HCV QS-RGQ Kit	COBAS AmpliPrep/ COBAS TaqMan HCV Test, v.2.0
Target sequence	5′NCR	5′NCR	5′NCR
Amplification method	qPCR	qPCR	qPCR
Detection method	Fluorescence	Fluorescence	Fluorescence
Internal control	Heterologous	Heterologous	Homologous
Standards	Four EQS	Four EQS	One IQS
Analytical measuring range (IU/ml)	$1.2 \times 10^1 - 1.0 \times 10^8$	$3.5 \times 10^1 - 1.8 \times 10^7$	$1.5 \times 10^1 - 1.0 \times 10^8$

EQS: external quantitation standards; IQS: internal quantitation standard.

Tab. 8.6: Comparison of currently used commercial assays for HCV genotyping.

Characteristics	Manufacturer and details			
	Abbott	Roche Molecular Diagnostics	Siemens	Siemens
Kit name	RealTime HCV Genotype II	Linear Array HCV Genotyping Test	Trugene HCV Genotyping Assay	VERSANT HCV Genotype 2.0 Assay
Target sequence	5′NCR, NS5B	5′NCR	5′NCR	5′NCR
Amplification method	qPCR	PCR	qPCR	PCR
Detection method	Fluorescence	Hybridization	Sequencing	Hybridization

particularly under antiviral therapy for HBV. It is required to monitor antiviral therapy (usually pegylated interferon-α) targeting HDV. Since the disease is relatively rare laboratory-developed PCR assays are commonly used. Several assays based on qPCR have been published but no preferred assay can be recommended.

HDV can be differentiated into distinct genotypes. Equal detection of all genotypes by the primers used as well as dissolving the tertiary RNA structure of HDV during RT require specific attention. Even though there are some indications that the genotypes may influence treatment outcome, genotyping is rarely done.

8.3.6 HEV

HEV RNA may be determined by conventional RT-PCR or qRT-PCR in patients with positive anti-HEV antibodies to confirm active infection. Detection of HEV RNA in stool specimens indicates infectivity; however, the virus is only detectable in stool for a short time period. Serum and plasma are thus the more common sample types used. Quantitation is not necessary because no specific antiviral treatment is available. Several home-brew molecular assays have been published. Recently, a commercial qPCR-based assay has been brought on the market (RealStar HEV RT-PCR Kit, altona Diagnostics) and further assays will follow in the near future.

8.3.7 Herpes viruses

Among the herpes viruses, not only EBV but also CMV, HSV, HHV6, and VZV should be taken into consideration as a cause of viral hepatitis, particularly under immunosuppression. Primary infections as well as reactivations can be accompanied by liver inflammation. Assays based on qPCR are well established for these persisting viruses because of their importance in the immunocompromised host (see also Chapter 9).

8.3.8 Yellow fever virus and hemorrhagic fever viruses

Yellow fever and dengue viruses are positive-strand RNA viruses, while hemorrhagic fever viruses (Lassa, Marburg, Ebola) are negative-strand RNA viruses. To confirm cases with uncertain serologic results or early stages of infection, conventional RT-PCR or RT-qPCR may be employed. Notably, those viruses (except dengue and yellow fever viruses) are graded as class 4 pathogens and samples from patients with the presumptive diagnosis may only be handled by adequately equipped laboratories.

8.4 Postanalytics – interpretation of results

Detection of nucleic acids from hepatitis viruses A to E proves active infection. If serum transaminase activity is elevated and/or histological evaluation of a liver biopsy indicates inflammatory liver disease, any of the hepatitis viruses A to E for which NAT was positive has to be regarded the causative agent. For HBV, HCV and HDV, clinical course and time course of positive testing determine whether it is considered an acute or a chronic infection.

For other viruses, liver inflammation, i.e. hepatitis, is only one of several clinical symptoms. Adenoviruses as well as herpes viruses may be detected during primary infection or during reactivation, which is mostly triggered by immunosuppression (see also Chapter 9).

8.4.1 HAV/HEV

The diagnosis of hepatitis A and hepatitis E relies on serological and/or molecular testing. The detection of IgM antibodies against HAV or HEV indicates an acute infection. This is confirmed by nucleic acid testing for HAV or HEV RNA in stool or blood indicating infectivity. A positive NAT result requires isolation of the patient or appropriate hygiene measures and close monitoring of liver function, especially in pregnant women and in patients with pre-existing liver disease. There is no specific antiviral therapy but supportive care may be necessary.

8.4.2 HBV

For HBV screening, serological testing for HBsAg and anti-HBc antibodies is recommended. Usually, patients positive for HBV DNA are also HBsAg positive. However, very recent infections may lack HBsAg detection. In addition, HBsAg-mutant viruses may escape serological detection as well as patients with ongoing anti-HBs seroconversion or occult hepatitis B (i.e. isolated seropositivity for anti-HBc). In these patients,

HBV DNA may be the only positive marker, although HBc antibodies usually develop later on.

Detection of HBsAg should be confirmed by neutralization assays and/or testing for HBV DNA. All patients positive for HBsAg and HBV DNA require periodical follow-up because they are at risk of developing liver disease including liver cirrhosis and hepatocellular carcinoma. However, there is no consensus about test intervals; they rely on the clinical presentation (degree of liver inflammation/liver damage) and on whether antiviral therapy is indicated or not.

All patients with chronic hepatitis B are candidates for antiviral therapy. According to the current EASL guidelines, treatment criteria for chronic hepatitis B include an HBV DNA level of greater than 2000 IU/ml (equivalent to approx. 1.0×10^4 copies/ml) and an ALT activity above the upper limit of normal. However, patients with advanced liver fibrosis or cirrhosis need antiviral therapy irrespective of the virus titer. HBV therapy should aim at eliminating the virus, but current antivirals only block reverse transcription and do not achieve hepatitis B cure.

Reactivation of HBV in patients who had controlled viremia before is frequently observed under immunosuppression and should be avoided by ongoing treatment with nucleos(t)ide analogs.

8.4.3 HDV

Testing for HDV co-infection is recommended if a new HBV infection has been diagnosed or if testing has not been done during the course of chronic hepatitis B. Superinfection with HDV should be excluded especially during unexplained exacerbations of hepatitis B. If specific HDV antibodies are detected, HDV PCR is required to determine the activity of virus replication because IgM testing lacks specificity. If antiviral therapy is indicated, quantitative PCR is done to monitor treatment success. The final goal is elimination of HDV RNA.

8.4.4 HCV

For HCV screening, serological testing for anti-HCV antibodies is performed. Because there may be unspecific reactions, testing for HCV RNA is required for confirmation. Usually, patients positive for HCV RNA also test positive for HCV antibodies. However, in immunocompromised patients and in the initial phase of an acute infection, antibodies may not yet be detectable. After clearance of HCV (persistently undetectable HCV RNA), HCV antibodies may also vanish. This may occur not only after antiviral therapy of persistent HCV infection, but also during spontaneous clearance of an acute infection.

Patients with chronic hepatitis C, especially those with elevated transaminases and/or liver fibrosis or extrahepatic manifestations, should be offered antiviral treatment, irrespective of the initial virus titer. The goal is to eliminate HCV (undetectable HCV RNA) and reduce HCV-associated morbidity and mortality. The chances for virus elimination steadily increase with novel DAAs being developed. Hepatitis C treatment is monitored and treatment duration is guided by quantitative determination of HCV RNA. Results are given in IU/ml relative to a genotype 1 WHO standard. However, between different molecular assays, there is considerable variation. To ensure virus elimination, ultrasensitive PCR assays with a limit of quantitation of at least 15 IU/ml are recommended. However, response to interferon-based therapy does not only rely on the virus but also on the host genotype (e.g. IL-28B single nucleotide polymorphism), which has to be determined by specific assays.

8.5 Take-home messages

- Hepatitis viruses A through E are hepatotropic viruses belonging to different virus families. They represent the most frequent causes of viral hepatitis.
- Although adenoviruses and herpes viruses show a different tropism but may also cause hepatitis, particularly in immunocompromised patients.
- Serologic tests are used for screening, while molecular assays are necessary to confirm positive serologic results, to determine active infection as well as infectivity, and to monitor response to treatment in chronic hepatitis B, C, and D infections.
- Molecular diagnostics of viral hepatitis largely relies on (RT-)qPCR, which allows quantitation of viral nucleic acids. For HAV and HEV, which usually do not become chronic, qualitative results are sufficient.
- To assess response to treatment in chronic hepatitis B and C, ultrasensitive (RT-)qPCR is recommended.

8.6 Further reading

[1] European Association for the Study of the Liver (2012) EASL clinical practice guidelines: management of chronic hepatitis B. J. Hepatol. 57:167–185.
[2] European Association for the Study of the Liver (2011) EASL clinical practice guidelines: management of chronic hepatitis C virus infection. J. Hepatol. 55:245–264.
[3] Lok, A.S., McMahon, B.J. (2009) Chronic hepatitis B: update 2009. Hepatology 50:661–662.
[4] Ghany, M.G., Strader, D.B., Thomas, D.L., Seeff, L.B. (2009) Diagnosis, management, and treatment of hepatitis C: an update. Hepatology 49:1335–1374.

9 Pathogens relevant in transplantation medicine
Marco Ciotti and Harald H. Kessler

Infectious complications are a major cause of mortality and morbidity in transplant recipients including bone marrow and solid organ transplant recipients due to immunosuppressive therapy. Several viruses including adenoviruses, cytomegalovirus (CMV), Epstein-Barr virus (EBV), human herpesvirus 6 (HHV-6), human herpesvirus 8 (HHV-8), varicella zoster virus (VZV), and the polyomaviruses BK (BKPyV) and JC (JCPyV) have been recognized as significant pathogens in transplantation medicine.

Adenoviruses may cause both localized and systemic disease in transplant recipients. Risk factors for primary infection or reactivation include younger age, allogeneic transplantation, T-cell depletion, unrelated or HLA-mismatched grafts, and low T-cell count.

The CMV is a ubiquitous beta-herpesvirus which infects up to 100% individuals, especially in the first two decades of life. In immunocompetent individuals, primary infection is usually asymptomatic or mild. CMV has the ability to establish lifelong latent infection following primary infection. Under certain conditions, CMV can reactivate, resulting in asymptomatic viral shedding or development of disease. Severe CMV disease is generally restricted to the immunocompromised or immunologically immature host. The CMV is thus the most important opportunistic viral pathogen for immunocompromised patients, with CMV seronegative transplant recipients at highest risk of developing CMV disease and its complications. Other risk factors include the immunosuppressive regimen (type of drug, dosage, timing, duration), host factors (age, comorbidity, leukopenia, lymphopenia, genetic factors). Furthermore, co-infections with HHV6 and HHV7 have been considered as risk factors. To avoid a lethal outcome of CMV disease, starting treatment at the earliest stage is of extreme significance. The level of CMV DNA has been found to be an important prognostic marker for the ongoing disease. Detection and quantitation of CMV DNA has been implemented and has been used together with pp65-antigenemia testing in the routine diagnostic laboratory.

The EBV is a gamma-herpesvirus which is a frequent cause of infection in humans showing a seroprevalence of 90–95% in adults worldwide. The virus remains latent in B-cell lymphocytes following primary exposure and thus has major importance in the immunocompromised patient. Quantitation of EBV DNA in these patients may provide information for initiating treatment and monitoring response to therapy. Information on quantitative viral load may guide a preemptive strategy to reduce the incidence and level of EBV reactivation in transplant recipients. EBV is associated with development of certain malignancies, including post-transplant lymphoproliferative disease (PTLD) and lymphomas. Quantitation of EBV DNA may thus not only reduce the incidence of EBV reactivation but also the subsequent development of PTLD and lymphomas. Indeed, a correlation between the EBV DNA level and the likelihood of development of PTLD exists. High EBV DNA levels in whole blood or plasma are seen in almost all patients with PTLD. Risk factors for PTLD are as follows: EBV seronegative status at time of transplant, active primary infection at time of transplant, patient age (children and older adults), immunosuppressive drug regimen and intensity, CMV and other viral co-infections.

The HHV-6 is a beta-herpesvirus. It has the ability to establish lifelong latency in mononuclear cells following primary exposure which usually occurs in the childhood. In contrast, primary infection in an adult seems to be a rare event. Chromosomally integrated HHV-6 (ciHHV-6) is found in about 1% of the population and the virus is vertically transmitted in a Mendelian mode. HHV-6 includes two molecularly and biologically distinct variants, A and B. While variant A has been associated with a case of syncytial-giant cell hepatitis, variant B is associated with unremarkable febrile illness or with roseola infantum also called exanthem subitum. In the immunosuppressed patient, HHV-6 may reactivate, usually within the first 4 weeks after transplantation, and produce systemic infections of the central nervous system (CNS) including meningitis and encephalitis, infections of the respiratory tract including pneumonia and hepatitis. In addition, HHV-6 infection increases the risk of CMV disease after kidney and liver transplantation. Although there are antiviral drugs available that inhibit the replication of HHV-6 *in vitro*, evaluation of such agents in larger clinical trials has not been done yet.

The HHV-8 is a member of the gamma-herpesvirus sub-family. It does not seem to be ubiquitous in the general population; the distribution appears related to behavioral and geographic factors. Infection with HHV-8 seems to occur most commonly through sexual transmission including saliva as a vehicle of transmission. Furthermore, HHV-8 may be transmitted through blood transfusion and transplantation. The virus has the ability to establish lifelong latency. Activation of HHV-8 replication in the latently infected cells is responsible for viral spread and presumed to contribute to the development of HHV-8 associated diseases including Kaposi's sarcoma, multicentric Castleman disease, primary effusion lymphoma and diffuse large B-cell lymphoma. Kaposi's sarcoma develops in 0.5–28% of solid organ transplant recipients depending on the patient's geographic origin and the immunosuppressive regimen used. In the majority of patients, the disease is a result of HHV-8 reactivation. The risk of developing Kaposi's sarcoma is greatest within the first 2 years following transplantation and decreases afterwards, probably due to the reduction of immunosuppressive therapy.

The VZV is an alpha-herpesvirus. After primary infection, VZV establishes a lifelong latency in dorsal root ganglia and cranial nerves. It may reactivate years to decades later as herpes zoster or "shingles". Decline in cell-mediated immunity is a risk factor for herpes zoster, and in transplant patients the incidence of herpes zoster can be more than 10 times that of the general population. Localized herpes zoster usually presents as a painful vesicular rash that involves ≤2 adjacent dermatomes. Secondary complications include bacterial superinfection and postherpetic neuralgia which occurs in approximately 20–40% of transplant recipients. VZV infections can also produce systemic infections of the CNS, respiratory tract, and hepatitis, mainly in the immunocompromised patient. Because the clinical presentation of VZV dermal disease can be confused with that produced by HSV, laboratory diagnosis is of major importance for distinguishing HSV from VZV infections. In the immunocompromised patient, quantitation of VZV DNA may provide information for initiating treatment and monitoring response to therapy. Information on quantitative viral load may guide a preemptive strategy to reduce the incidence and level of VZV reactivation in transplant recipients.

The BKPyV was originally isolated from the urine of a renal graft recipient with the initials B.K. This human polyomavirus infects most people subclinically during childhood. Approximately 80% of adults worldwide are found to be seropositive. After primary infection, the virus remains latent in the kidney and reactivation may occur through immunosuppression. In kidney transplant patients, BKPyV reactivation can cause an inflammatory interstitial nephritis referred to as polyomavirus-associated nephropathy (PyVAN) that can lead to graft loss. Nephritis induced by BKPyV occurs usually within the first year after transplantation. However, in up to 25% of renal graft recipients, it may start later. In bone marrow recipients, high BKPyV DNA load in urine has been associated with polyomavirus-associated hemorrhagic cystitis (PyHC) which develops in about 5–40% of patients with BKPyV viruria.

The JCPyV was isolated by cell culture from brain tissue of a patient with the initials J.C. suffering from progressive multifocal leukoencephalopathy (PML). JCPyV infection occurs in 35–70% of the general population. After primary exposure, the virus persists in the renourinary tract. JCPyV has been linked to PML in severely immunocompromised patients such as AIDS patients and to cases of meningitis or encephalitis in patients without PML. Occasionally, JCPyV has been recognized as another etiological agent of PyVAN.

Further viruses such as enteroviruses, the human herpesvirus 7 (HHV-7), herpes simplex viruses type 1 and 2 (HSV-1 and HSV-2), parvovirus B19 may also be relevant. In the immunocompromised patient, enteroviruses may be responsible for severe systemic disease with meningitis, meningoencephalitis, and myocarditis (see also Chapter 12). The HHV-7 shows a very similar clinical presentation in comparison with HHV-6. In immunosuppressed patients, virus reactivation with interstitial pneumonia and organ rejection has been observed. Complications due to virus reactivation are also the major problem with HSV-1 and HSV-2. Herpes simplex viruses may cause conjunctivitis, herpes simplex dermatitis, eczema herpeticum, generalized HSV infection with hepatitis or pneumonia, meningitis, and herpes encephalitis (see also Chapter 12). While parvovirus B19 infection is usually mild in immunocompetent people, immunocompromised patients may develop chronic infection with severe anemia and thrombocytopenia. Furthermore, meningitis, encephalopathy, myocarditis, vasculitis, and hepatitis may be associated with parvovirus B19 infection.

9.1 Clinical manifestations

Areas of infection and clinical manifestations associated with infections in transplant recipients are shown in Tab. 9.1.

9.2 Preanalytics

Specimens useful for detection of nucleic acids may be collected from different body sites depending on the symptoms and the suspected diagnosis (Tab. 9.2). Notably, there is currently no consensus on the optimal blood compartment for routine molecular herpesvirus DNA testing. The most meaningful results may be obtained by quantitation of EBV DNA in EDTA whole blood including both the cellular and the cell-free compartments from serial or sequential specimens obtained from the same patient. However, herpesvirus DNA can be detected in peripheral blood mononuclear cells during both active disease and latency. When using a quantitative molecular assay, it must be taken into consideration that a cut-off to distinguish active infection from latent infection has not yet been identified. Furthermore, the comparability of results obtained from serial or sequential specimens may be impaired through fluctuating leukocyte counts.

Tab. 9.1: Virus related areas of infection and clinical manifestations.

Virus	Area of infection	Clinical manifestation
Adenovirus	Respiratory tract, gastrointestinal tract, liver, bladder, eyes, CNS	Pneumonia, pharyngitis, gastroenteritis, hepatitis, hemorrhagic cystitis, keratoconjunctivitis, encephalitis, disseminated disease
CMV	Bone marrow, respiratory tract, gastrointestinal tract, liver, retina, CNS	Myelosuppression, pneumonia, hepatitis, gastroenteritis, colitis, retinitis, encephalitis
EBV	B cells, lung, CNS, bone marrow	Post-transplantation lymphoproliferative disease, lymphomas, graft-versus-host disease, pneumonia, encephalitis, anemia, thrombocytopenia
HHV-6	Liver, gastrointestinal tract, bone marrow, lung, CNS	Hepatitis, gastroduodenitis, colitis, myelosuppression, pneumonia, encephalitis
HHV-8	Endothelia	Kaposi's sarcoma, Castleman's disease, primary effusion lymphoma
VZV	Dermatome, CNS, respiratory tract	Chickenpox, zoster, encephalitis, pneumonia
BKPyV	Bladder, kidney	Hemorrhagic cystitis, polyomavirus associated nephropathy
JCPyV	CNS, kidney	Progressive multifocal leukoencephalopathy, meningitis, encephalitis, polyomavirus associated nephropathy

Tab. 9.2: Virus related specimens useful for detection of nucleic acids.

Virus	Specimen
Adenovirus	EDTA whole blood, EDTA plasma, nasopharyngeal swab or aspirate, throat washing, bronchoalveolar lavage, stool, gastric or colon biopsies, urine, cerebrospinal fluid, conjunctival swab
CMV	EDTA whole blood, EDTA plasma, cerebrospinal fluid, throat washing, bronchoalveolar lavage, aqueous humor, bone marrow, stool, urine, amniotic fluid
EBV	EDTA whole blood, EDTA plasma, cerebrospinal fluid, bronchoalveolar lavage, bone marrow
HHV-6	EDTA plasma, serum, EDTA whole blood, cerebrospinal fluid, bronchoalveolar lavage, bone marrow
HHV-8	EDTA whole blood, EDTA plasma, serum, pleural effusion, ascites
VZV	EDTA whole blood, EDTA plasma, swab, cerebrospinal fluid, bronchoalveolar lavage
BKPyV	EDTA plasma, EDTA whole blood, urine
JCPyV	Cerebrospinal fluid, EDTA whole blood, EDTA plasma, urine

9.2.1 Adenoviruses

EDTA whole blood or plasma samples are usually employed for detection of adeno-virus DNA in transplant recipients. In the case of localized disease, alternative specimens such as nasopharyngeal aspirate or swab, throat washing, bronchoalveolar lavage, stool, gastric or colonic biopsies, cerebrospinal fluid, conjunctival swab, or urine may be collected.

9.2.2 CMV

There is currently no consensus on the optimal blood compartment for routine CMV DNA testing. EDTA whole blood, peripheral blood leukocytes, and plasma have been used for routine diagnosis. Because results obtained from different sample materials may be incomparable, a single specimen type should be used for monitoring a single patient. In several studies, EDTA whole blood has been found to be superior to peripheral blood leukocytes or plasma. With a highly sensitive molecular assay, latent virus may be detected but the detection of low-level CMV DNA in EDTA whole blood seems neither to be clinical relevant nor a better predictor of CMV viremia or disease. In absence of clear cutoffs for introduction and discontinuation of antiviral therapy, ultrasensitive molecular assays may result in overtreatment with increased toxicity for the patient and increased cost. Further specimen types for detection of CMV DNA in the immunocompromised patient include cerebrospinal fluid, throat washing, bronchoalveolar lavage, stool, urine, aqueous humor, and bone marrow.

9.2.3 EBV

Presently, there is no consensus on the optimal blood compartment for routine molecular EBV DNA testing. In transplant recipients, both EDTA whole blood and plasma specimens appear to be informative. EDTA plasma should be considered the specimen of choice in cytopenic patients. In patients with encephalitis, detection of EBV DNA in cerebrospinal fluid is considered the standard method.

9.2.4 HHV-6

In case of suspected HHV-6 active infection, detection of HHV-6 DNA is performed in EDTA plasma or serum, while detection in peripheral blood mononuclear cells is of limited clinical relevance because of virus persistence. Quantitative PCR performed on whole blood can detect up to 5.5 log/ml of HHV-6 DNA because of chromosomal integration. This has sometimes lead to unnecessary treatment on the assumption that this high viral load reflected active infection. The role of chromosomally integrated HHV-6 after transplantation is currently unknown. Preliminary results suggest that transplant patients with integrated HHV-6 are more likely to experience bacterial infection and GVHD. For CNS disease, detection of HHV-6 DNA in cerebrospinal fluid specimens is the standard method. However, false-positive results may be obtained if mononuclear cells are present in the CSF. For respiratory HHV-6 infection, detection of HHV-6 DNA in bronchoalveolar lavages is meaningful. Association with interstitial pneumonitis has been reported.

9.2.5 HHV-8

The most meaningful results for routine molecular HHV-8 DNA testing may be obtained by quantitation of viral DNA in EDTA whole blood. During infection, viral DNA can be detected in both peripheral blood mononuclear cells and plasma. Recently, it has been suggested that peripheral blood mononuclear cells and plasma are both adequate specimen types for quantitation of HHV-8 DNA. In patients seropositive for HHV-8 or receiving an organ from a seropositive donor, monitoring of HHV-8 viremia after transplantation might be a useful strategy to determine the risk of developing Kaposi sarcoma. However, the frequency and duration of testing and the level of clinically relevant replication that predicts the risk of disease have not been determined yet. Nevertheless, quantitative PCR has been suggested to assess the response to treatment in patients with Kaposi sarcoma. Further studies are needed to evaluate the clinical utility of these approaches. Quantitative PCR may also have a role in the diagnosis and monitoring of multicentric Castleman disease because high HHV-8 viremia can be detected at disease onset and used as disease-associated marker during follow-up. Finally, PCR testing of pleural effusion or ascites may be useful for a definitive diagnosis of primary effusion lymphoma.

9.2.6 VZV

In the immunocompromised patient, quantitation of VZV in EDTA whole blood as well as in plasma is clinically relevant. For suspected dermal VZV infection, detection of VZV DNA in swabs is the standard method. For CNS disease, detection of VZV DNA in CSF specimens is the standard method. For respiratory VZV infection, detection of VZV DNA in bronchoalveolar lavages is meaningful.

9.2.7 BKPyV

The specimens of choice are EDTA plasma, EDTA whole blood, and urine. Quantitation of BKPyV DNA in EDTA plasma was shown to have the highest predictive value for PyVAN. A persisting plasma BKPyV DNA level exceeding 4 $\log_{10}$ copies/ml for more than 3 weeks is highly suggestive of PyVAN and evaluation with graft biopsy is recommended.

9.2.8 JCPyV

Cerebrospinal fluid (CSF) is the specimen of choice for the diagnosis of PML, meningitis, and encephalitis. The performance of the JCPyV PCR assay is of paramount importance for CSF testing. The amount of JCPyV DNA in CSF may be related to the level of viral replication in brain lesions and to factors such as the clearance rate from CSF, the size and the anatomical location of the brain lesion, and the recovery of general and specific immune functions. In the laboratory confirmed diagnosis of PML, the LLQ of the PCR assay used should be not higher than 50 copies/ml. In less than 1% of kidney transplant patients, JCPyV has been recognized as the etiological agent of PyVAN. In these patients, very high urine JCPyV DNA loads exceeding 7 $\log_{10}$ copies/ml were found along with urinary shedding of "decoy cells" and cytopathic epithelial cells with characteristic intranuclear inclusions. In contrast to urine, JCPyV viremia is usually low or undetectable and transient. Of course, the involvement of BKPyV must be excluded.

9.3 Analytics

9.3.1 Sample preparation

For a number of years, manual nucleic acids extraction protocols based on phenol–chloroform extraction have been used successfully in molecular diagnostic laboratories around the world. However, manual methods have several drawbacks (see also Chapter 4). Therefore, sample preparation protocols on automated platforms have largely replaced manual protocols.

9.3.2 Nucleic acids amplification and detection

Today, qPCR has largely replaced conventional PCR and become the method of choice for amplification, detection, and quantification of adenovirus, polyomavirus and herpesvirus DNAs in transplant recipients. Quantitative results are usually reported as copies/ml.

9.3.2.1 Adenoviruses

For the detection and quantitation of adenovirus DNA, laboratory-developed assays are commonly used. qPCR assays usually utilize primers targeting the fiber or hexon genes. To guarantee detection of 56 serotypes identified so far, an extended BLAST analysis is strongly recommended. Furthermore, subsets of primers or a single primer pair in combination with locked nucleic acid nucleotides for improved hybridization and melting point analysis should be employed. A summary of currently frequently used commercial assays in Europe is provided in Tab. 9.3.

9.3.2.2 CMV

Following the era of laboratory-developed assays based on conventional PCR in the nineties, several assays based on qPCR have been established more recently. Molecular assays based on quantitative qPCR have been shown to provide several important advantages to detection of CMV antigen even though some studies indicated general agreement between the two methods. Advantages of molecular assays based on quantitative qPCR include increased sensitivity for early detection of CMV infection

Tab. 9.3: Comparison of currently frequently used commercially available assays for the quantitation of adenovirus DNA.

Characteristics	Manufacturer and details		
	Altona Diagnostics	**bioMerieux/Argene**	**Nanogen/ELITechGroup**
Kit name	RealStar Adenovirus PCR Kit 1.0	Adenovirus R-gene	Adenovirus ELITe MGB Kit
Target sequence	NA	Hexon gene	Hexon gene
Amplification method	qPCR	qPCR	qPCR
Detection method	Fluorescence	Fluorescence	Fluorescence
Internal control	Heterologous	Heterologous	Heterologous
Standards	Four EQS	Four EQS	Four EQS
Analytic measuring range	4.0×10^1–4.0×10^7 copies/µl	5.0×10^2–1.0×10^7 copies/ml	2.5×10^2–2.5×10^7 genome equivalents/ml (cellular samples); 1.5×10^2–1.5×10^7 genome equivalents/ml (noncellular samples)

NA: data not available; EQS: external quantitation standards.

or reactivation, utility for patients with neutropenia, wide range of linearity (up to 8 $\log_{10}$), ability to process a large number of specimens and the potential for increased accuracy of results through precision instrumentation.

However, laboratory-developed molecular assays for the detection of CMV DNA are almost unique for each laboratory and usually lack standardization. More than ten different target regions of the CMV genome and at least three different units of result reports have been described. In this context, it must be emphasized that the choice of the target region requires special attention. In several laboratory-developed molecular assays, the glycoprotein B gene (UL55) has been used as the target region. However, CMV variants may not be quantified using quantitative qPCR targeting the glycoprotein B gene. Sequence analysis revealed a single base pair mutation in the target sequence of the down-stream probe. More seriously, several additional sequence differences with the probes used in the glycoprotein B gene assay do exist.

Several commercial qPCR-based assays have been developed for the quantitation of CMV DNA. A summary of currently frequently used commercial assays in Europe is provided in Tab. 9.4. All of those assays show a wide range of linearity and several of them have been evaluated including comparisons with detection of CMV antigen and other molecular assays and have been found to be suitable for the detection and quantitation of CMV DNA in the routine diagnostic laboratory.

Recently, the 1st WHO International Standard for Human Cytomegalovirus (HCMV) for Nucleic Acid Amplification Techniques (NIBSC 09/162) was introduced. Results obtained by standardized assays for quantitation of CMV DNA should thus be reported in IU/ml.

9.3.2.3 EBV

Several gene targets have been selected for laboratory-developed assays for the quantitation of EBV DNA based on qPCR including genes coding for DNA polymerase, thymidine kinase, Epstein-Barr nucleic antigen type 1, glycoprotein B, ZEBRA protein and a nonglycosylated membrane protein. Similar to the detection of CMV, the choice of the target region requires attention as EBV sequence variation is a recognized phenomenon. Especially, if the latent membrane protein genes are used as the target region, sequence variation may lead to an underestimation of the EBV viral load or even to a false-negative result.

Several commercial assays have been developed for the quantitation of EBV DNA. A summary of currently frequently used commercial assays in Europe is provided in Tab. 9.5. All of those assays are based on qPCR. Recently, evaluation studies on commercially available molecular assays for the quantitation of EBV DNA have been published. Assays investigated were found to be reliable and suitable for routine EBV-associated disease monitoring. However, it must be mentioned that one of those assays (the LightCycler EBV Quant Kit) amplifies part of the EBV latent membrane protein 2 gene. In the presence of the recognized EBV sequence variation, a shifted melting curve may result because of an altered binding of one of the fluorescent

Tab. 9.4: Comparison of currently frequently used commercially available assays for the quantitation of CMV DNA.

Characteristics	Manufacturer and details				
	Abbott	bioMerieux/Argene	Nanogen/ELITechGroup	Qiagen	Roche
Kit name	RealTime CMV	CMV R-gene	CMV ELITe MGB Kit	artus CMV QS-RGQ Kit	COBAS AmpliPrep/COBAS TaqMan CMV Test
Target sequence	UL34 (early protein) and pUL80.5 (scaffolding protein)	UL83 (lower matrix phosphoprotein 65)	UL123 (major immediate early protein)	UL123 (major immediate early protein)	UL54 (DNA polymerase)
Amplification method	qPCR	qPCR	qPCR	qPCR	qPCR
Detection method	Fluorescence	Fluorescence	Fluorescence	Fluorescence	Fluorescence
Internal control	Heterologous	Heterologous	Heterologous	Heterologous	Homologous
Standards	Two EQS	Four EQS	Four EQS	Four EQS	One IQS
Analytic measuring range	3.1×10^1–1.6×10^8 IU/ml (plasma); 6.2×10^1–1.6×10^8 IU/ml (EDTA whole blood)	5.0×10^2–1.0×10^7 copies/ml	3.2×10^2–2.5×10^7 genome equivalents/ml (EDTA whole blood)	7.9×10^1–1.0×10^8 copies/ml (plasma); 1.0×10^3–5.0×10^7 copies/ml (EDTA whole blood)	1.5×10^2–1.0×10^7 copies/ml (plasma)

EQS: external quantitation standards; IQS: internal quantitation standard

Tab. 9.5: Comparison of currently frequently used commercially available assays for the quantitation of EBV DNA.

Characteristics	Manufacturer and details				
	bioMerieux/Argene	diagenode	Nanogen/ELITechGroup	Qiagen	Roche
Kit name	EBV R-gene	EBV Realtime PCR	EBV ELITe MGB Kit	artus EBV QS-RGQ Kit	LightCycler EBV Quant Kit
Target sequence	BXLF1 (thymidine kinase)	BNRF1 (nonglycosilated membrane protein)	EBNA-1 (Epstein-Barr nucleic antigen 1)	EBNA-1 (Epstein-Barr nucleic antigen 1)	LMP2 (latent membrane protein 2)
Amplification method	qPCR	qPCR	qPCR	qPCR	qPCR
Detection method	Fluorescence	Fluorescence	Fluorescence	Fluorescence	Fluorescence
Internal control	Heterologous	Heterologous	Heterologous	Heterologous	Heterologous
Standards	Four EQS	Four EQS	Four EQS	Four EQS	Four EQS
Analytic measuring range	$5.0 \times 10^2 - 1.0 \times 10^7$ copies/ml	$8.2 \times 10^2 - 1.0 \times 10^7$ copies/ml	$2.5 \times 10^2 - 2.5 \times 10^7$ genome equivalents/ml (cellular samples); $1.3 \times 10^2 - 1.3 \times 10^7$ genome equivalents/ml (noncellular samples)	$3.2 \times 10^2 - 1.0 \times 10^7$ copies/ml (plasma)	$1.0 \times 10^3 - 2.0 \times 10^7$ copies/ml (plasma); $2.0 \times 10^3 - 2.0 \times 10^7$ copies/ml (EDTA whole blood)

EQS: external quantitation standards.

hybridization probes. Therefore, careful analysis of the melting curve is of paramount importance to avoid false-negatives when using this assay.

While qPCR assays for quantitation of EBV DNA show a high level of reproducibility within individual laboratories, a substantial variability was observed between laboratories. The poor interlaboratory reproducibility contributes to the lack of consensus on the viral load threshold that should trigger a therapeutic intervention. It is thus important to monitor an individual patient at a single site. The recent introduction of the 1st WHO International Standard for Epstein-Barr Virus for Nucleic Acid Amplification Techniques (NIBSC 09/260) will contribute to the standardization of EBV DNA quantitation. Results obtained by standardized assays for quantitation of EBV DNA should thus be reported in IU/ml.

9.3.2.4 HHV-6

Laboratory-developed molecular assays based on qPCR have been established and used for the detection of HHV-6 DNA in plasma. There are just a few commercial assays on the market presently (Tab. 9.6). Recently, one of the assays, the CMV HHV6,7,8 -R-gene test, was found useful for the routine diagnostic laboratory when compared with two HHV-6 in-house quantitative qPCR methods.

9.3.2.5 HHV-8

Laboratory-developed molecular assays based on qPCR have been established and used for the quantitation of HHV-8 DNA. Primers and probes located in the ORF26 region of the viral genome, coding for the HHV8 minor capsid protein, have been employed. There are only two commercial assays on the market presently (Tab. 9.6).

9.3.2.6 VZV

Several laboratory-developed assays for the qualitative detection/quantitation of VZV DNA based on qPCR have been established. Gene targets include genes coding for DNA polymerase, glycoprotein B, DNA binding protein and immediate-early transactivator.

A number of commercial assays have been developed for the qualitative detection/quantitation of VZV DNA. A summary of currently frequently used commercial assays in Europe is provided in Tab. 9.7. All of those assays are based on qPCR; however, no evaluation studies have been published so far.

9.3.2.7 BKPyV

Laboratory-developed assays for the detection and quantitation of BKPyV are increasingly replaced by commercially available kits. qPCR assays usually utilize primers targeting the early gene region of the virus. A summary of currently frequently used commercial assays in Europe is provided in Tab. 9.8.

Tab. 9.6: Comparison of currently frequently used commercially available assays for the quantitation of human herpesvirus 6 (HHV-6) and 8 (HHV-8) DNA.

Characteristics	Manufacturer and details			
	bioMerieux/Argene	Altona Diagnostics	Nanogen/ELITechGroup	Nanogen/ELITechGroup
Kit name	CMV HHV6,7,8 R-gene	RealStar® HHV-6 PCR Kit 1.0	HHV6 ELITe MGB Kit	HHV8 ELITe MGB Kit
Target sequence	U57 (major capsid protein) for HHV6; ORF26 (minor capsid protein) for HHV8	NA	U67 (ORF 13R)	ORF26 (minor capsid protein)
Amplification method	qPCR	qPCR	qPCR	qPCR
Detection method	Fluorescence	Fluorescence	Fluorescence	Fluorescence
Internal controls	Heterologous	Heterologous	Heterologous	Heterologous
Standards	Four EQS	Four EQS	Four EQS	Four EQS
Analytic measuring range	5.0×10^2–1.0×10^7 copies/ml (HHV-6); 5.0×10^2–1.0×10^7 copies/ml (HHV-8)	1.0×10^1–1.0×10^8 copies/µl	2.5×10^2–2.5×10^7 genome equivalents/ml (cellular samples); 1.5×10^2–1.5×10^7 genome equivalents/ml (noncellular samples)	2.5×10^2–2.5×10^7 genome equivalents/ml (cellular samples)

NA: data not available; EQS: external quantitation standards.

Tab. 9.7: Comparison of currently frequently used commercially available assays for the qualitative detection/quantitation of varicella zoster virus (VZV) DNA.

Characteristics	Manufacturer and details				
	bioMerieux/Argene	Cepheid	Nanogen/ELITechGroup	Qiagen	Roche
Kit name	HSV1 HSV2 VZV R-gene	Affigene VZV tracer	VZV ELITe MGB Kit	artus VZV QS-RGQ Kit	LightCycler VZV Qual Kit
Target sequence	ORF17 (tegument host shut-off protein)	ORF62 (immediate early protein)	ORF29 (major DNA binding protein)	ORF38 (virion protein)	ORF29 (major DNA binding protein)
Amplification method	qPCR	qPCR	qPCR	qPCR	qPCR
Detection method	Fluorescence	Fluorescence	Fluorescence	Fluorescence	Fluorescence
Internal control	Heterologous	Heterologous	Heterologous	Heterologous	Heterologous
Standards	Four EQS	–[a]	Four EQS	Four EQS	–[a]
Analytic measuring range	5.0×10^2–1.0×10^7 copies/ml	–[b]	2.5×10^2–2.5×10^7 genome equivalents/ml (cellular samples); 1.0×10^2–1.2×10^7 (noncellular samples)	5.0×10^2–1.0×10^8 copies/ml	–[c]

[a] Qualitative assay.
[b] Limit of detection 7.2×10^1–3.3×10^2 copies/ml (depending on the sample preparation protocol employed).
[c] Limit of detection 1.5×10^2 copies/ml (CSF), 3.8×10^2 copies/ml (swab).
EQS: external quantitation standards.

9.3.2.8 JCPyV

A few commercial assays are available for detection and quantitation of JCPyV. Primers/probes usually target the Large T antigen gene. A summary of available assays is reported in Tab. 9.9.

Tab. 9.8: Comparison of currently frequently used commercially available assays for the quantitation of BKPyV DNA.

Characteristics	Manufacturer and details				
	Altona Diagnostics	bioMerieux/ Argene	Cepheid	Nanogen/ ELITechGroup	Qiagen
Kit name	RealStar BKV PCR Kit 1.0	BK Virus R-gene	Affigene BKV trender	BKV ELITe MGB Kit	*artus* BK Virus QS-RGQ Kit
Target sequence	NA	t-Ag (small t antigen)	VP1 gene (late protein)	T-Ag (Large T antigen)	VP2-VP3 genes (late proteins)
Amplification method	qPCR	qPCR	qPCR	qPCR	qPCR
Detection method	Fluorescence	Fluorescence	Fluorescence	Fluorescence	Fluorescence
Internal control	Heterologous	Heterologous	Heterologous	Heterologous	Heterologous
Standards	Four EQS	Four EQS	Two EQS	Four EQS	Four EQS
Analytic measuring range	1.0×10^1–1.0×10^9 copies/µl	5.0×10^2–1.0×10^{11} copies/ml	5.0×10^2–1.0×10^8 copies/ml (EDTA whole blood); 1.0×10^2–1.0×10^8 copies/ml (serum)	1.5×10^2–1.5×10^7 genome equivalents/ml	5.0×10^1–9.3×10^7 copies/ml (plasma); 1.0×10^2–1.0×10^9 (urine)

NA: data not available; EQS: external quantitation standards.

Tab. 9.9: Comparison of currently frequently used commercially available assays for the quantitation of JCPyV DNA.

Characteristics	Manufacturer and details		
	Altona Diagnostics	Eurospital Diagnostic	Nanogen/ELITechGroup
Kit name	RealStar JCV PCR Kit 1.0	Euro-RT JCV Kit	JCV ELITe MGB Kit
Target sequence	NA	T-Ag (Large T antigen)	T-Ag (Large T antigen)
Amplification method	qPCR	qPCR	qPCR
Detection method	Fluorescence	Fluorescence	Fluorescence
Internal control	Heterologous	Heterologous	Heterologous
Standards	Four EQS	Five EQS	Four EQS
Analytic measuring range	1.0×10^1–1.0×10^9 copies/µl	1.0×10^2–1.5×10^6 copies/µl	6.0×10^2–6.0×10^7 genome equivalents/ml (noncellular samples)

NA: data not available; EQS: external quantitation standards.

9.4 Postanalytics – interpretation of results

9.4.1 Adenoviruses

For adenoviruses, no viral DNA load cut-off value that may predict prognosis or disease progression has been defined so far. It is thus advisable to monitor the viral DNA load for the particular patient over time, rather than refer to an absolute value. It is important to keep in mind that asymptomatic reactivation of adenovirus may occur in solid organ transplant recipients but may not be associated with disease.

Transplanted children are particularly at risk of developing adenovirus infection. There exists an increased risk of developing disseminated disease if the adenovirus DNA load exceeds 1×10^5 copies/ml. Furthermore, in hematopoietic stem cell recipients, the virus may be detected in blood 2–3 weeks before the onset of symptoms offering the opportunity for intervention. Similar observations have been reported for children after solid organ transplantation. However, a few patients may also clear the virus spontaneously without treatment. Anyway, qPCR may detect early infection and identify patients at risk through frequent determination of the viral load. However, there is presently no consensus on the threshold viral load value relevant for treatment.

9.4.2 CMV

There exists a clear relationship between high CMV DNA levels in blood and the clinical progression of CMV infection not only in solid organ but also in bone marrow recipients. Monitoring the viral load at biweekly intervals within the first month post-transplant, at weekly intervals up to three months post-transplant, and at monthly intervals up to one year post-transplant allows identification of patients at risk for developing CMV disease. Both the CMV DNA load at the onset of active infection and the increase of viral load correlate with CMV disease in transplant recipients. This information is essential for the initiation of (preemptive) antiviral therapy whose efficacy is best monitored through quantitative qPCR. However, it must be stated that a clear-cut threshold value regarding prediction of disease progression, initiation of antiviral therapy, and response to treatment has not yet been defined. The availability of the first WHO International Standard for CMV will contribute to the standardization of assays for quantitation of CMV DNA and facilitate the definition of thresholds. If the CMV DNA load increases continuously during antiviral therapy, nucleotide mutations may be present and thus be responsible for the lack of therapeutic efficacy. In this case, sequencing of the target gene (especially UL97 and UL54) should be done for confirmation or exclusion.

9.4.3 EBV

The introduction of qPCR for detection of EBV DNA in plasma or whole blood has improved the management of transplant recipients. While low EBV DNA levels may vanish without treatment, high EBV DNA levels obtained from consecutive samples strongly indicate development of complications including PTLD requiring immediate treatment. The median onset of PTLD is 2 months after bone marrow transplantation and 6 months after solid organ transplantation. EBV DNA monitoring at biweekly intervals within the first month post-transplant, at weekly intervals up to three months post-transplant, and at monthly intervals up to one year post-transplant allows the identification of patients at high risk of developing PTLD. The availability of the first WHO International Standard for EBV will contribute to the standardization of assays for quantitation of EBV DNA.

9.4.4 HHV-6

Detection of HHV-6 DNA in blood correlates well with active replication and quantitation of HHV-6 DNA by qPCR may allow discrimination between latent and active infection but a reference value to be used as threshold value has not yet been defined. Nevertheless, reactivation of HHV-6 in the post-transplantation period is characterized by a progressive increase of viral load. A biphasic progress of the viral load may be observed with an increase within 4 weeks after transplantation, followed by a decrease afterwards. Finally, HHV-6 levels should be similar to those observed prior to transplantation. Complications such as thrombocytopenia, graft-versus-host-disease, anemia, leucopenia, and encephalitis may occur in patients with a high HHV-6 load. Therefore, sequential measurement of the HHV-6 load is advisable to prevent viral complications.

9.4.5 HHV-8

Quantitation of the HHV-8 load in blood by qPCR assays may be useful for monitoring of post-transplant patients. There exists a significant correlation between viral load and disease progression. However, with regard to the development of Kaposi's sarcoma, no HHV-8 DNA load cut-off has been defined.

9.4.6 VZV

Quantitation of VZV-DNA levels has been a valuable diagnostic approach in immunosuppressed patients. Although no consensus was reached on VZV-DNA load threshold values, several studies suggested the viral load as a marker for the severity of the disease and a predictor of outcome. High viral load is associated with clinical symptoms and poor clinical outcome. It was reported that immunosuppressed patients with >1.0 x 10^5 VZV DNA copies/ml in the peripheral blood showed organ involvement such as pneumonia, hepatitis, and generalized infection and some of those patients died. In addition, detection of VZV DNA in CSF correlates strongly with meningitis and encephalitis. Frequent monitoring of VZV DNA by quantitative qPCR is essential to recognize primary infections and reactivations and to monitor patients undergoing antiviral therapy.

9.4.7 BKPyV

Reactivation of BKPyV is a serious problem in transplant recipients. Because detection of BKPyV DNA in plasma precedes the onset of PyVAN, determination of the viral load in urine and plasma may help to identify patients at risk of developing PyVAN in renal allograft recipients. These patients should thus be monitored 3-monthly within the first two years post-transplant and annually thereafter up to the fifth year post-transplant. In case of a positive result, reduction of immunosuppressive therapy should strongly be considered. Preemptive reduction of immunosuppression has been reported to lead to viral clearance and to rescue the transplant. In hematopoietic stem cell recipients, hemorrhagic cystitis may develop. After detection of BKPyV DNA in the urine, the patient must be monitored at weekly intervals.

9.4.8 JCPyV

High JCPyV viral load in urine was associated with the development of PyVAN in kidney transplant patients which had been found BKPyV-negative. No general monitoring recommendation for JCPyV DNA exists presently. Nevertheless, persisting "decoy cells" shedding or decline in renal function should prompt qPCR on JCPyV DNA in urine after exclusion of BKPyV presence. Treatment of PyVAN through reduction of immunosuppressive therapy can stabilize allograft function and rescue the organ.

9.5 Take-home messages

- Recent advances in molecular technologies have greatly improved virological diagnosis of pathogens relevant in transplantation medicine.
- Modern automated sample preparation allows the introduction of a large variety of specimens including EDTA whole blood, plasma, serum, tissue, cerebrospinal fluid, throat washing, bronchoalveolar lavage, urine, and stool.

- qPCR is used to monitor the viral load in transplant recipients. It allows detection of the onset of disease at a very early stage and is essential for both the initiation and the monitoring of preemptive therapy if available.
- Presently, no clear-cut threshold values regarding prediction of disease progression, initiation of antiviral therapy, and response to treatment exist due to the lack of standardization of molecular assays.

9.6 Further reading

[1] Anderson, E.J. (2008) Viral diagnostics and antiviral therapy in hematopoietic stem cell transplantation. Curr. Pharm. Des. 14:1997–2010.

[2] Espy, M.J., Uhl, J.R., Sloan, L.M., Buckwalter, S.P., Jones, M.F., Vetter, E.A., Yao, J.D.C., Wengenack, N.L., Rosenblatt, J.E., Cockerill, F.R. III, Smith, T.F. (2006) Real-Time PCR in clinical microbiology: applications for routine laboratory testing. Clin. Microbiol. Rev. 19:165–256.

[3] Gulley, M.L., Tang, W. (2010) Using Epstein-Barr viral load assays to diagnose, monitor, and prevent posttransplant lymphoproliferative disorder. Clin. Microbiol. Rev. 23:350–366.

[4] Hirsch, H.H., Brennan, D.C., Drachenberg, C.B., Ginevri, F., Gordon, J., Limaye, A.P., Mihatsch, M.J., Nickeleit, V., Ramos, E., Randhawa, P., Shapiro, R., Steiger, J., Suthanthiran, M., Trofe, J. (2005) Polyomavirus-associated nephropathy in renal transplantation: interdisciplinary analyses and recommendations. Transplantation 79(Suppl 10):1277–1286.

[5] Lee, S.O., Brown, R.A., Razonable, R.R. (2011) Clinical significance of pretransplant chromosomally integrated human herpesvirus-6 in liver transplant recipients. Transplantation 92:224–229.

[6] Lisboa, L.F., Åsberg, A., Kumar, D., Pang, X., Hartmann, A., Preiksaitis, J.K., Pescovitz, M.D., Rollag, H., Jardine, A.G., Atul Humar, A. (2011) The clinical utility of whole blood versus plasma cytomegalovirus viral load assays for monitoring therapeutic response. Transplantation 91:231–236.

10 Pathogens in lower respiratory tract infections

Margareta Ieven and Katherine Loens

Lower respiratory tract infections (LRTIs) are an important problem. They occur frequently, are associated with significant morbidity and mortality, are present in a variety of healthcare settings, and impose a considerable cost to European healthcare services.

In developed countries, LRTIs remain a leading cause of death. WHO statistics estimate that mortality from respiratory infections is 48/100,000 worldwide and ranges from 40 to 50/100,000 in Europe according to the WHO annual report at http://www.who.org. These figures stress the importance of early recognition of those patients who are severely ill or at risk of becoming severely ill. Of all respiratory tract infections (RTIs), approximately one-third are thought to involve the lower respiratory tract, with approximately 10% community-acquired pneumonia (CAP), and the remaining two-thirds affect the upper respiratory tract.

Presently, there is still a great deficit in the etiologic diagnosis of community-acquired lower respiratory tract infections (CA-LRTIs); in most studies, more than 50% of cases remain without an etiologic diagnosis resulting in unnecessary or inappropriate antibiotic prescribing.

10.1 Clinical importance of different etiologic agents

Wide variations between studies in the frequency of each microorganism can be explained by several factors including differences in study populations (e.g. age range, risk factors), geographical area, samples studied, and microbiological methods. For example, studies focus on bacterial agents or viruses or intracellular bacteria.

In the majority of studies on LRTIs, there are a large proportion of cases with no pathogen identified because the appropriate tests were not performed (as is usually the rule in outpatients) or the organism was missed. Most studies in mild infections suggest that microbial etiologies in outpatients are similar to those of hospitalized patients.

Streptococcus (S.) pneumoniae is the most frequent microbial agent found in bacteriologically documented pneumonia in both elderly and middle-aged adults. Of all etiologic agents detected in LRTIs and CAP, *S. pneumoniae* is responsible for more than 20–30%, both in the community and in hospitalized patients. The risk of pneumonia due to *S. pneumoniae* is higher in old people than in the general population. The occurrence of pneumonia due to *S. pneumoniae* is more frequent in institutionalized than in non-institutionalized people of the same age, and the relative risk of pneumococcal infection increases linearly with age.

Among atypical infections, *Mycoplasma (M.) pneumoniae* is the most common, followed in frequency by *Legionella* and *Chlamydophila* species. *M. pneumoniae* belongs to the class of the Mollicutes; it has been associated with a wide variety of acute and chronic diseases. RTIs with *M. pneumoniae* occur worldwide and in all age groups. The proportion of LRTIs in children and adults including CAP associated with *M. pneumoniae* infection has ranged from 0% to >50% during the past 10 years, varying with age and geographic location of the population examined and the diagnostic methods used. The true role of *M. pneumoniae* in RTIs remains a challenge, given the wide variation of data from studies with equally wide variation of and lack of standardized diagnostic methods.

Legionella (L.) pneumophila is an aerobic, nutritionally fastidious Gram-negative rod. Since the organism was first identified in 1976 during an outbreak at an American Legion Convention in Philadelphia, *Legionella* has been recognized as a relatively common cause of both community-acquired and hospital-acquired pneumonia, with *L. pneumophila* causing >90% of the cases of Legionnaires' disease (LD). LD is a form of bacterial pneumonia that is characterized by fever, chills, and a dry cough associated with muscle aches and occasional diarrhea. *Legionella* is commonly found in cooling towers, humidifiers, and potable water distribution systems. Severity ranges from mild to fatal, with an average mortality of 15% observed in immunocompromised patients.

Chlamydophila (C.) pneumoniae, an obligate intracellular bacterium, has been associated with a variety of diseases but most importantly with respiratory infections that occur worldwide and in all age groups. Studies of *C. pneumoniae* published since the 1990s mention the organism to be associated in <5% to up to >40% of LRTIs in both children and adults varying with age, geographical location, the population studied and particularly the diagnostic methods used.

Bordetella (B.) pertussis, a small Gram-negative bacterium, is responsible for the vast majority of pertussis or "whooping cough" cases; *B. parapertussis* causes a mild pertussis-like illness. Other species, including *B. bronchiseptica*, may infrequently cause respiratory infections in humans, mostly in patients who are immunocompromised. In the present vaccine era, the overall incidence of pertussis has been reduced dramatically. Although pertussis has generally been considered an infection of children associated with considerable morbidity and mortality, it is being increasingly seen over the past two decades in adults. The increase in reported pertussis over the past two decades is mainly due to a greater awareness of pertussis and perhaps to the use of several less effective vaccines. Studies of prolonged cough illnesses in adolescents and adults reveal that 13–20% could be a result of *B. pertussis* infection. Serologic studies suggest that the rate of *B. pertussis* infection in adolescents and adults is approximately 2.0% per year.

The importance of viruses as causal agents has been confirmed in LRTI. In CAP, the most common etiologic agents after *S. pneumoniae* are respiratory viruses which are involved in 5–20% of cases. A number of reports have appeared on the epidemiology of viruses in acute respiratory infection (ARI) but most are restricted to a few viruses (influenza viruses, sometimes together with respiratory syncytial virus, rhinoviruses, metapneumovirus, and coronaviruses). Great variations occur in the time, place, and age groups studied.

Influenza viruses consist of three different types: A, B, and C. Among influenza A viruses, distinct subtypes determined by the hemagglutinin (HA) and neuraminidase (NA) genes can be distinguished. Influenza A causes seasonal epidemics with the epidemic period varying mostly between 3 and 8 weeks. Pandemics can occur after the emergence of a novel virus subtype because of genetic reassortment of HA and NA genes occurring when two different subtypes co-infect the same host. Influenza is a febrile illness characterized by a sudden increase in temperature, headache, malaise, myalgia, and respiratory symptoms such as sore throat and cough. During the seasonal epidemics, influenza infections are responsible for a significant number of hospitalizations each year: especially the elderly over the age of 65 are most at risk, followed by young children under the age of 5.

Respiratory syncytial virus (RSV) is the most common cause of viral LRTIs in very young children and is the major cause of bronchiolitis and pneumonia in infants below 2 years of age. Two subtypes can be distinguished: groups A and B, of which RSV A seems to be responsible for the more severe infections. Disease often begins with upper respiratory symptoms but rapidly progresses to bronchiolitis or pneumonia with wheezing and respiratory distress. Not only in very young children but also in elderly RSV is commonly associated with LRTIs and pneumonia. In Europe, peak prevalence of RSV infections is usually observed in winter between December and March.

Human parainfluenzaviruses (HPIVs) are divided into four serotypes (HPIV-1 to HPIV-4) that are able to cause respiratory infections in humans. Especially HPIV-1 is the most important etiologic agent in croup in very young children but can also be responsible for milder respiratory tract infections. HPIV-2 has also been associated with croup but less frequently than HPIV-1. HPIV-3 is associated with severe LRTIs. HPIV-4 is the least common virus of this group. HPIVs occur at any time of the year but small outbreaks can be observed in autumn, mostly due to HPIV-1, and spring, mostly due to HPIV-3.

Rhinoviruses, with more than 200 serotypes, have until recently only been associated with the common cold. However, they are increasingly important, now being associated with asthma exacerbations in both children and adults, wheezing, acute respiratory distress, serious LRTIs and even CAP in children, elderly, and immunocompromised patients.

Adenoviruses, including more than 51 serotypes, can produce a variety of infections from respiratory infections to conjunctivitis, keratoconjunctivitis, and gastroenteritis. The most common serotypes involved in human respiratory infections are serotypes 1–5, causing mostly mild symptoms but more severe disease such as bronchiolitis or even pneumonia may also occur occasionally, especially in pediatric transplant patients.

Coronaviruses are also more prevalent than previously thought. Furthermore, the importance of more recently discovered viruses such as human bocavirus, human metapneumovirus (hMPV), and new coronaviruses is becoming more evident. Based on symptoms, hMPV infection is very similar to RSV infection both causing upper respiratory tract infection (URTI) and/or LRTI, especially bronchiolitis but also pneumonia, most frequently in very young children. The peak incidence of HMPV infections usually follows RSV but both peaks overlap frequently. Although the role of some of these new viruses becomes clearer in specific patient populations, more studies are needed to identify the clinical relevance of some others such as bocavirus.

As diagnostic techniques improve, it also becomes clear that multiple organisms may be found. Pediatric studies found polymicrobial infections in CAP: dual viral infection was present in 0–14%, dual bacterial infection in 0–14%, and mixed viral–bacterial infection in 3–30%. In hospitalized adult non-immunocompromised patients, polymicrobial CAP occurs in up to 6% at all ages and as well in inpatients and outpatients; frequent combinations are bacterial with an atypical organism or a virus. In general, *S. pneumoniae* is the most prevalent microorganism.

10.2 Specimen collection

Specimens useful for detection of etiologic agents in LRTI can be collected from different body sites, especially in cases of suspected pneumococcal pneumonia, or can be different types of specimens from the respiratory tract. It should be mentioned here that for a number of respiratory etiologies there is currently no consensus on the optimal sampling strategy. Specimens which may be used for PCR-based tests or PCR combined with other detection methods are presented in Tab. 10.1.

10.2.1 *S. pneumoniae*

Blood cultures are still considered the gold standard for culture-based diagnosis of pneumococcal pneumonia, while blood specimens are still far from being used for PCR-based diagnosis. However, recent studies show that the detection of bacterial DNA load in whole blood supports the diagnosis of *S. pneumoniae* infection in patients with CAP. High levels of *S. pneumoniae* DNA may be useful for severity assessment. Systematic comprehensive studies need to be performed to confirm the clinical usefulness of PCR on plasma or whole blood.

The specificity of respiratory specimens for the diagnosis of bacterial pathogens, either by culture or by amplification-based techniques, is not high in LRTIs because

Tab. 10.1: Comparison of specimens and sampling methods in different studies for the detection of different respiratory pathogens.

Pathogen	Specimen	Method
Streptococcus pneumoniae	Sputum	RQ PCR
Mycoplasma pneumoniae	Sputum > TW > NPS > OPS	PCR
	OPS > NPS, NPA	PCR
	OPS > BAL > sputum	PCR
	Sputum > NPA	PCR, EIA
	Sputum > OPS	PCR, NASBA, Gen-Probe test, culture
	NPS = OPS	PCR
Chlamydophila pneumoniae	NPS > TS	PCR, culture
	Sputum > NPS = OPS	PCR, culture
	OPS > NPS	PCR
	Sputum > NPA > OPS	PCR
	NPS > OPS > sputum	PCR
Legionella pneumophila	BAL, sputum	PCR
Bordetella pertussis	NPA, NA, sputum	PCR
Influenza virus A/B	(N)PFS = NPA	PCR, DFA
	NPS > NPA > OPS	PCR, culture, DIF, Binax Now, Directigen Flu A+B
RSV	NPA > NPFS	PCR, DIF
	PFS = NPA > NS > OPS	PCR
All viruses	NPA > NS > OPS	PCR, culture, DIF
	NFS = NA > unpreserved saline	PCR

BAL: bronchoalveolar lavage; NA: nasal aspirate; NASBA: nucleic acid sequence-based amplification; NPA: nasopharyngeal aspirate; NPFS: nasopharyngeal flocked swab; NFS: nasal flocked swab; NPS: nasopharyngeal swab; NS: nasal swab; OPS: oropharyngeal swab; PCR: polymerase chain reaction; PFS: pernasal flocked swab; TW: throat wash.

of contamination with the upper airway flora. Several techniques have been proposed to achieve accurate discrimination between colonization and infection. Diagnostic accuracy is improved by the use of bronchoalveolar lavage (BAL) specimens in severely ill hospitalized patients.

It is generally accepted that sputum is the best unsterile respiratory specimen for the recovery of *S. pneumoniae*. Sputum specimens must be representative of lower respiratory secretions. The most widely used method to assess the acceptability in this regard is based on cytologic criteria. The specimen should therefore be screened by microscopic examination for the relative number of polymorphonuclear cells and squamous epithelial cells in the lower power (10) field. Only valid specimens ($\leq$10 squamous epithelial cells and $\geq$25 polymorphonuclear cells/field) should be examined further. It may be difficult to obtain good quality, purulent sputum. Many pneumonia patients do not produce sputum, particularly older patients and children. Inhalation of hypertonic saline may be helpful.

Although increasing information is becoming available on the use of PCR for *S. pneumoniae* in respiratory samples, the value of the Gram stain on sputum specimens should not be underestimated. It has been suggested that a properly collected and read Gram stain provides a simple, readily available, rapid, and inexpensive test result. It thus can be a reliable test for the early etiologic diagnosis of bacterial pneumonia in bacteremia patients.

10.2.2 *M. pneumoniae, L. pneumophila*, and *C. pneumoniae*

Because of its fastidious nature, *M. pneumoniae* is not routinely cultured from respiratory specimens. Most studies are PCR-based on both LRT and URT specimens. Different specimens have been used such as sputum, nasopharyngeal swabs (NPS) or oropharyngeal swabs (OPS) or washes, BAL, or pleural fluid. It has been demonstrated in several studies that sputum is superior compared with other respiratory specimens for the molecular detection of *M. pneumoniae*. The high diagnostic sensitivity of sputum PCR can be explained by the higher number of *M. pneumoniae* organisms in the pulmonary alveoli than on the epithelium of the URT, which has been demonstrated in experimentally infected hamsters. Consequently, if a sputum sample is available, it might be the best specimen for *M. pneumoniae* detection. A nasopharyngeal aspirate (NPA) or OPS might be the second best option for analysis by nucleic acid amplification techniques (NAATs).

In patients with CA-LRTI caused by *L. pneumophila*, bronchoscopically taken specimens such as BAL are the preferred specimens in severely ill patients admitted to the hospital. Sputum specimens are generally considered the best alternative specimens for the isolation of *L. pneumophila* in non-severely ill patients. Combining test results from more than one site appears to improve the diagnostic accuracy.

The choice of the respiratory specimen may also have a major impact on the sensitivity of *C. pneumoniae* isolation and PCR. Sputum or an NPS may be the preferred specimen for detection of *C. pneumoniae* by NAATs.

10.2.3 *B. pertussis*

The NPA is the best sample for infants; it offers superior sensitivity compared with swabs. NPSs provide valid specimens from older children, adolescents, and adults. Sputum samples or throat washes may represent an alternative for adolescents and adults. The sensitivity of detecting *Bordetella* DNA in these materials must be verified because, for culture, throat washes were found to be less suitable.

10.2.4 Respiratory viruses

In patients with viral respiratory infections, recovery of specimens must be done in the early stage of infection. Specimens are obtained from respiratory epithelium (nose, trachea, bronchi) and placed into a viral transport medium. Bronchoalveolar lavage may be useful in immunocompromised patients.

The optimal specimen is the nasopharyngeal aspirate (NPA) obtained by slight suction on a catheter introduced consecutively in both nostrils, if necessary after introduction of 1 ml of physiological saline through the catheter. Alternatively, an NPS may be collected. Different types of swabs and transport media have been used. For virus transport systems, no standard has been defined. The Copan system (Copan, Brescia, Italy) combining flocked swabs and universal transport medium (UTM) for collection and transport is a universal system compatible with antigen detection kits, direct fluorescent antibody (DFA) testing, culture, and PCR. With the new flocked swabs, a significantly higher number of epithelial cells are collected, providing a better specimen quality. It has been shown that a higher number of positives are found with NPSs collected with flocked swabs in comparison to NPSs collected with Dacron swabs.

10.3 Diagnostic procedures

A wide variety of diagnostic procedures and techniques are applied for the detection of the etiologic pathogens of CA-LRTI. Traditional diagnostic culture methods lack sensitivity, are not feasible in many contexts, and focus only on a few of the many etiologic agents. In recent years, several previously unknown respiratory agents were discovered whose *in vitro* culture is very slow or even unrealized: hMPV, the novel coronaviruses NL63 and HKU1, human bocavirus, the new polyomaviruses, and mimivirus. For the so-called "atypical" bacterial pneumoniae caused by *M. pneumoniae*, *L. pneumophila*, and *C. pneumoniae*, and for *B. pertussis* infection, traditional diagnostic methods are also too insensitive and too slow, producing a result only after several days.

Over the past two decades, NAATs, particularly PCR and nucleic acid sequence-based amplification (NASBA), have revolutionized the diagnostic procedures for the management of patients with RTI, resulting from a combination of improved sensitivity and specificity, the potential for automation, and the production of rapid results within a few hours, leading to superior clinically useful microbiologic information.

10.3.1 Sample preparation and nucleic acid extraction

Respiratory samples are among the most difficult clinical specimens. Nucleic acid extraction originally performed with phenol-chloroform has been widely replaced by the Boom method and by commercial sample preparation kits. Conventional manual nucleic acid extraction from clinical samples is the most labor-intensive and critical part in nucleic acid diagnostic assays. Methods are time consuming, labor intensive and susceptible to contamination. The probability of false-positive results because of contamination increases with the number of manipulations involved in sample processing. Automated nucleic acid extraction systems show high flexibility in type and numbers of samples which can be handled. With a wide range of sample input and elution volumes and a short turnaround time, they provide another opportunity to optimize NAATs for clinical services. Automated sample preparation performed by instruments such as MagNA Pure and NucliSens easyMAG show equal or better and more consistent performance than manual techniques. An advantage of the NucliSens easyMAG instrument is that it can be applied for a broad range of different specimens such as blood, sputum, serum, and throat swabs and both DNA and RNA can be isolated in the same run. Furthermore, nucleic acid extracts can be used in combination with different amplification methods including PCR and NASBA.

10.3.2 Amplification and detection methods for individual agents

For all etiologic agents described, an overview of possible targets combined with single target amplification techniques is presented in Tab. 10.2.

Tab. 10.2: Mono nucleic acid amplification based tests (PCR-based unless other technique mentioned).

Pathogen	Target	Detection procedure
Mycoplasma pneumoniae	P1 gene	Agarose gel electrophoresis and hybridization, Light Cycler, I-Cycler, TaqMan, LAMP
	16S rRNA gene, *tuf* gene, ATPase gene	Agarose gel electrophoresis and hybridization
	16S rRNA	NASBA and electrochemiluminescence detection, real-time NASBA, Qβ-replicase
	CARDS toxin gene	ABI Prism 7500
Chlamydophila pneumoniae	16S rRNA	NASBA and electrochemiluminescence detection, real-time NASBA
	16S rRNA gene	Agarose gel electrophoresis and hybridization, EIA, Light Cycler, TaqMan
	MOMP gene	Agarose gel electrophoresis and hybridization or RFLP, Light Cycler, TaqMan, ABI Prism 7700
	MOMP mRNA	NASBA and electrochemiluminescence detection
	pmp4 gene	Light Cycler
	60-kDa protein gene	Agarose gel electrophoresis and RFLP, EIA
	DnaK gene, KDO transferase mRNA	Agarose gel electrophoresis and hybridization
	53 kDa gene	Agarose gel electrophoresis
	Cloned *PstI* fragment	Agarose gel electrophoresis and hybridization, Light Cycler

(Continued)

Tab. 10.2: (*Continued*)

Pathogen	Target	Detection procedure
Legionella pneumophila	*mip* gene, 5S rRNA gene	Agarose gel electrophoresis and hybridization, Light Cycler
	16S rRNA gene	Agarose gel electrophoresis and hybridization, Light Cycler, ICT-strip
	16S rRNA	NASBA and electrochemiluminescence detection, real-time NASBA
	*gyr*B gene	Microarray
	*dot*A gene	Agarose gel electrophoresis and hybridization, ABI prism 7000
	*dna*J gene	ICT-strip
	*flh*A gene	Fluorescence depolarization
Bordetella pertussis	IS*481*	Agarose gel electrophoresis and hybridization or RFLP, Light Cycler
	Pertussis toxin promoter	Agarose gel electrophoresis and hybridization, LAMP
	Pertactin gene	TaqMan
	BP*3385*	ABI Prism 7900HT
	BP*283*, BP*485*	Light Cycler
	Cya gene, Porin gene	Agarose gel electrophoresis and hybridization
Influenza A/B	HA gene	LAMP
	Matrix gene	Smart Cycler
	NP gene	Real-time NASBA
	M gene	LAMP, agarose gel electrophoresis
RSV	F1 fusion protein gene	Agarose gel electrophoresis, multi-component nucleic acid enzymes technology
	N gene	EIA, TaqMan, Light Cycler, AmpliTaq, ABI Prism 7900HT, agarose gel electrophoresis
	1B gene	EIA
	L gene	Agarose gel electrophoresis and RFLP, Light Cycler
	F gene	Light Cycler, TaqMan, ABI Prism 7900HT, real-time NASBA
	Matrix protein gene	TaqMan
	RNase P gene	TaqMan
Parainfluenzavirus 1, 2, 3	Hexon gene, F gene	Agarose gel electrophoresis and hybridization
	HN gene	NASBA and electrochemiluminescence detection
	HA-NA	EIA
	pol gene	TaqMan
	L-gene	TaqMan
Adenovirus	H gene	Agarose gel electrophoresis + hybridization or RFLP, Light Cycler, TaqMan, hydrolysis probes
	VA region	Agarose gel electrophoresis and RFLP
	Long fiber gene	Agarose gel electrophoresis and hybridization

(*Continued*)

Tab. 10.2: (*Continued*)

Pathogen	Target	Detection procedure
Human metapneumovirus	N gene	EIA, Light Cycler, ABI Prism 7000
Coronaviruses	Polymerase gene	Agarose gel electrophoresis
	RNA polymerase gene	Agarose gel electrophoresis, Light Cycler, TaqMan, I-Cycler
	N-gene	Agarose gel electrophoresis, Light Cycler, TaqMan, I-Cycler, ABI Prism 7700
	BNI fragment	Light Cycler, BioSystems 7000 SDS
	Nucleocapsid gene	Light Cycler, TaqMan, I-Cycler, LAMP
	P-gene, M-gene	I-Cycler
Rhinoviruses	5-NCR	Agarose gel electrophoresis and hybridization, Light Cycler, TaqMan
	5-NCR	NASBA and electrochemiluminescence detection
Bocavirus	NS1, NP gene	TaqMan, hydrolysis probes

EIA: enzyme immunoassay; ICT: immunochromatography; LAMP: loop-mediated isothermal amplification method; NASBA: nucleic acid sequence-based amplification.

10.3.2.1 *S. pneumoniae*

S. pneumoniae is the most common cause of CAP in adults and infants. Detection by conventional culture is cheap and facilitates antibiotic resistance determination. However, it is time-consuming and can be indeterminate, especially when antibiotics were administered prior to sampling. Several PCRs have been employed with varying degrees of success, using primers specific to repetitive regions and genes encoding rRNA, the pneumococcal surface adhesion A molecule (*psa*A), pneumolysin (*ply*), and the penicillin-binding protein. According to the guidelines of the Joint Taskforce of the European Respiratory Society and the European Society for Clinical Microbiology and Infectious Diseases, qualitative nucleic acid amplification tests for *S. pneumoniae* on pleural fluid, peripheral blood, and sputum add little to the existing diagnostic tests in sputum because they do not allow distinguishing colonization from infection.

In recent years, qPCR has improved diagnostics since this approach may help to distinguish carriage from infection. In a recent prospective study, qPCR was evaluated on sputum samples from patients with CAP admitted to the hospital. In patients with proven pneumococcal etiology, the yield from qPCR was almost twice as high as that from sputum culture. This figure suggests that in hospital-treated CAP patients, qPCR on sputum samples is a more sensitive method to detect *S. pneumoniae* than sputum culture and the previously chosen cutoff level corresponding to 10^5 CFU/ml was confirmed. Especially, when antibiotic treatment had been initiated, qPCR, together with urine antigen detection, was the best method to identify *S. pneumoniae*. The detection of *S. pneumoniae* by qPCR in plasma is also useful for the rapid detection of bacteraemic pneumococcal pneumonia. Detection of bacterial DNA load in whole blood supports the diagnosis of *S. pneumoniae* infection in patients with CAP. Bacterial load is associated with the likelihood of death, the risk of septic shock, and the need for mechanical ventilation. High genomic bacterial load for *S. pneumoniae* may be a useful tool for severity assessment.

Nevertheless, the identification of *S. pneumoniae* is rather complicated. Strains that are genotypically closely related to oral streptococci harbor the genes encoding the *S. pneumoniae* virulence factors *lyt*A and *ply*. PCR methods based on the *ply* gene are thus not specific and show high detection rates in saliva from culture-negative healthy individuals. Recently, new *S. pneumoniae* specific targets, Spn9802 and *lyt*A, have been reported.

10.3.2.2 *M. pneumoniae*

Serological methods, in particular the complement fixation test (CFT) and enzyme immunoassays (EIAs), are most widely used to diagnose *M. pneumoniae* infection. The application of PCR is increasingly accepted as a rapid diagnostic test because culture is too slow and too insensitive to be therapeutically relevant. None of the currently available NAATs has been extensively evaluated against reference culture. The sensitivity of NAATs is usually superior to that of traditional procedures; NAATs are thus increasingly considered as the "new gold standard". In-house and commercially available NAATs using a great variation of methods, including different targets (P1 gene, 16S rRNA, ATPase gene, *par*E gene, *tuf* gene; monoplex versus multiplex targets), different amplification methods (conventional, nested, real-time; RNA vs. DNA targets; PCR, NASBA), and different detection formats (agarose gel electrophoresis, SYBR green, TaqMan probe, hybridization probes, molecular beacons, microchip electrophoresis) have been described.

The availability of very sensitive NAATs allows a better interpretation of serological test results. The low incidence of IgM antibodies in acute phase serum specimens and the importance of the delay between two serum samples have been shown. However, data obtained from different studies indicate that no single test may be sensitive enough for the identification of *M. pneumoniae* in CAP. A combination of serology and PCR appears to be the most reliable approach for identification of *M. pneumoniae* in CAP.

10.3.2.3 *C. pneumoniae*

The combination of PCR and serum IgM detection seems to be preferable for the diagnosis of acute LRTI caused by *C. pneumoniae*.

For PCR, the MOMP gene and 16S rRNA gene are often used as targets. However, none of the available methods has been standardized, which has resulted in a wide variation of inter-laboratory test performance. In an effort to standardize diagnostic assays for *C. pneumoniae*, the US Centers for Disease Control and Prevention (CDC) published recommendations for diagnostic testing.

10.3.2.4 *L. pneumophila*

Serologic tests, although often used, never provide an early diagnosis of legionellosis and thus are rather an epidemiological than a diagnostic tool. In contrast, NAATs enable specific amplification of minute amounts of *Legionella* DNA and have the potential to detect infections caused by any *Legionella* species or serogroup. Diagnostic PCR assays usually target specific regions within the 16S rRNA gene, the 23S-5S spacer region, 5S rDNA, or the *mip* gene. Thus far, encouraging results obtained mostly from *in vivo* evaluations and small patient series have been reported. A commercial test (BD ProbeTec ET *L. pneumophila*; Becton Dickinson) that detects *L. pneumophila* serotypes 1–14 in sputum has been cleared by the FDA, but published data on performance characteristics are lacking.

The added value of qPCR for diagnosis of LD in routine clinical practice was evaluated in several clinical studies. Patients were evaluated if, in addition to PCR, the results of at least one of the following diagnostic tests were available: (1) culture for *Legionella* spp. on specific media; (2) detection of *L. pneumophila* antigen in urine specimens. An estimated sensitivity and specificity of 86% and 95% were found for the 16S rRNA-based PCR, and corresponding values of 92% and 98%, respectively, were found for the mip gene-based PCR. The added value of a *L. pneumophila*-specific PCR to a urinary antigen test was especially shown in patients with suspected Legionnaires' disease who produce sputum.

Assays based on qPCR were also shown to be useful for the diagnosis of *L. pneumophila* in CAP cases, although they suffered from a lower sensitivity in comparison to the urinary antigen test. Both qPCR and antigen testing should thus be considered as diagnostic tools for detection of legionellosis. High bacterial loads determined by qPCR in LRT samples were found to be useful for predicting disease severity, which may be an advantage of qPCR and warrant further investigation.

10.3.2.5 *B. pertussis*

The use of *B. pertussis*-specific PCR in combination with single-serum serology has been shown to increase the sensitivity for pertussis diagnosis. Furthermore, PCRs are more sensitive than culture for the detection of *B. pertussis* and *B. parapertussis*, especially in the late stage of the disease and after antibiotic treatment has started; qPCR formats are widely used now. Similar to culture, the sensitivity of PCR decreases with the duration of cough; however, due to its higher sensitivity, it may be a useful tool for diagnosis not only for the first 3–4 weeks of coughing, but even longer. A frequently used target for PCR detection of *B. pertussis* is the repetitive element IS481, which is absent in *B. parapertussis* but found in approximately 50 to a few hundred copies *B. pertussis*. Although IS481 is generally regarded as specific for *B. pertussis*, few studies detected it in *B. bronchiseptica*, also. Although infections with *B. bronchiseptica* are primarily associated with disease in mammals other than humans, human illness and carriage is increasing, particularly in infants and immunocompromised hosts with exposure to carrier animals. Although rare, disease caused by *B. bronchiseptica* in immunocompetent adults can also occur suggesting that PCR targeting IS481 might not be sufficiently specific for reliable identification of *B. pertussis*. Additionally, there has been concern about the specificity of detection of *B. pertussis* owing to sequence identity with *B. holmesii*. Although the pertussis toxin operon is present in *B. pertussis*, *B. parapertussis*, and *B. bronchiseptica*, the pertussis toxin promoter is a target for *B. pertussis*-specific assays using qPCR. However, it is consistently less sensitive than IS481. Some reference laboratories are using both targets.

10.3.2.6 Respiratory viruses

Antigens of the most common respiratory viruses such as influenza viruses, RSV, adenoviruses, and parainfluenzaviruses can be detected by direct immunofluorescence (DIF) or by commercially available EIAs. The sensitivities of these tests vary from 50% to >90% but the sensitivity of the DIF test is lower in adults and older persons than in children. Culture procedures for viruses and fastidious bacteria are too insensitive and are superseded by NAATs, preferably in real-time format and in combination with automated nucleic acid extraction. These approaches allow a result to be obtained within 4–5 h.

After conventional PCR, detection of amplification products is usually done by agarose gel electrophoresis with or without hybridization with a labeled probe, occasionally by restriction fragment length polymorphism. Alternatively, amplification products may be captured onto a solid phase and detected by an enzyme immunoassay that is more convenient for the examination of clinical samples in batches. Both NASBA and reverse transcriptase PCR have their advantages. One advantage of NASBA compared with PCR is that it is a continuous, isothermal process. In addition, RNA is the genomic material of numerous respiratory viruses. NAATs are available for all respiratory agents. If different viruses or serotypes in a family or a genus must be detected, consensus primers able to detect all these viruses can be used, but their ability to amplify all viruses with the same efficiency must be carefully evaluated. Through judicious choice of primers, the high specificity of NAATs can be ascertained. Targets and methods for conventional and real-time in-house developed NAATs for the detection of a single agent are presented in Tab. 10.2. Multiplex protocols are used for those virus groups that contain multiple subtypes. Several amplification protocols were developed to cover particular types or groups among adenoviruses. More than 200 rhinovirus types and their close relationship with enteroviruses constitute a special challenge. The suitable choice of primers and particularly of the hybridization probes should ensure a satisfactory coverage of the rhinovirus types. Face-to-face comparisons of two amplification protocols comparing two different conventional molecular amplification techniques applied to a considerable number of clinical specimens are rare.

Following the 2009 influenza A pandemic, innovative techniques for detection of influenza infection were introduced including the "simple amplification-based assay" (SAMBA) and the RAZOR EX thermocycler. The SAMBA Flu duplex test is a dipstick-based molecular assay developed to detect influenza A and B viruses intended for point-of-care diagnosis. The test utilizes isothermal amplification and visual detection of nucleic acids on a test strip. The entire test procedure (extraction, amplification, and detection) is performed on a semi-automated platform. The field-compatible, portable RAZOR EX instrument can be used for the rapid, near-patient detection of influenza viruses in clinical specimens. With this portable thermocycler, shipping of the sample to a diagnostic laboratory may be avoided.

10.3.3 Multiplex NAATs

Respiratory viruses and other so-called "atypical bacteria" are all responsible for RTIs that can produce clinically similar manifestations. In order to reduce expenses and hands-on time, both in-house and commercially available multiplex NAATs (MX-NAATs) for the simultaneous detection of up to 20 different respiratory pathogens including the "atypical" bacteria *M. pneumoniae*, *C. pneumoniae*, and *L. pneumophila* and respiratory viruses with a mixture of primers have been developed. Tab. 10.3 and Tab. 10.4 summarize various combinations of etiologic agents targeted in both

Tab. 10.3: In-house multiplex nucleic acid amplification based tests for detection of respiratory pathogens.

Year of introduction	Technique	Detection procedure	Pathogens targeted
2003	MX-PCR	Real-time	*B. pertussis, B. parapertussis*
	MX-PCR	Real-time	*M. pneumoniae, C. pneumoniae, L. pneumophila*
2004	MX-PCR	Microchip electrophoresis	*M. pneumoniae, C. pneumoniae, L. pneumophila*
	MX-PCR	Agarose gel electrophoresis	Influenza A, B and C, RSV A/B, parainfluenza 1–4, 2 coronaviruses, rhinovirus, enterovirus, adenovirus in two multiplex reactions
	MX-PCR	Agarose gel electrophoresis	*M. pneumoniae, C. pneumoniae,* influenza A and B, RSV, parainfluenza 1 and 3, human metapneumovirus, adenovirus, rhinovirus and enterovirus in two MX and one mono reaction

(Continued)

Tab. 10.3: (*Continued*)

Year of introduction	Technique	Detection procedure	Pathogens targeted
2005	Mass Tag MX-PCR	Masscode Tag	Influenza A and B, RSV A and B, parainfluenza 1–3, metapneumovirus, three coronaviruses, *C. pneumoniae*, *M. pneumoniae*, *L. pneumophila*, enterovirus, adenovirus
	MX-PCR	Real-time	Influenza A, B and C, RSV, parainfluenza 1–4, two coronaviruses, rhinovirus in three MX reactions
	MX-PCR	Real-time	Influenza A and B, RSV A/B, parainfluenza 1–3, human metapneumovirus, three coronaviruses, rhinovirus in four MX reactions
	MX-PCR	Agarose gel electrophoresis	*M. pneumoniae*, *C. pneumoniae*, *L. pneumophila*, *B. pertussis*
	MX-PCR	Agarose gel electrophoresis	*S. pneumoniae*, *H. influenzae*, *M. pneumoniae*, *C. pneumoniae*
2006	MX-PCR	Agarose gel electrophoresis	*B. pertussis*, *B. parapertussis*
	MX-PCR	Agarose gel electrophoresis	*M. pneumoniae*, *C. pneumoniae*, *B. pertussis*, *B. parapertussis*
2007	MX-PCR	Agarose gel electrophoresis	*M. pneumoniae*, *C. pneumoniae*, *L. pneumophila*, adenovirus
	MX-PCR	Resequencing microarray	*S. pneumoniae*, *M. pneumoniae*, *C. pneumoniae*, *S. pyogenes*, *Y. pestis*, *B. anthracis*, *F. tularensis*, influenza A and B, RSV A and B, parainfluenza 1, 3, adenovirus, two coronaviruses, rhinovirus, Lassa virus, vaccinia virus, ebola virus, variola major virus
	MX-PCR	Microarray	*M. pneumoniae*, *C. pneumoniae*, *B. pertussis*, *S. pyogenes*, influenza A and B, RSV, parainfluenza 1–3, adenovirus, three coronaviruses
2008	MX-PCR	Real-time	*B. pertussis*, *B. parapertussis*
	MX-PCR	Real-time	*M. pneumoniae*, *C. pneumoniae*
	MX-PCR	Real-time	*M. pneumoniae*, *C. pneumoniae*, *Legionella* spp.

(*Continued*)

Tab. 10.3: (*Continued*)

Year of introduction	Technique	Detection procedure	Pathogens targeted
	MX-PCR	Reverse line blot hybridization	*S. pneumoniae, H. influenzae, M. catarrhalis, S. aureus, M. pneumoniae, C. pneumoniae, L. pneumophila, B. pertussis, S. pyogenes, K. pneumoniae, M. tuberculosis*
	MX-PCR	Microarray	*S. pneumoniae, M. pneumoniae, C. pneumoniae, L. pneumophila, L. micdadei, B. pertussis, S. aureus*, influenza A and B, RSV A and B
	MX-PCR	Real-time	*Legionella* spp., *L. pneumophila*
	MX-PCR	Real-time	*B. pertussis, B. parapertussis, B. bronchiseptica, B. holmesii*
	MX-PCR	Reverse dot-blot	Influenza A and B, parainfluenza 1 and 3, RSV, rhinovirus, coxsackievirus
	MX-PCR	Real-time	Influenza A and B, parainfluenza 1–3, RSV A and B, adenovirus, rhinovirus, enterovirus, coronavirus OC43, human metapneumovirus, bocavirus
2009	MX-PCR	Agarose gel electrophoresis	Adenovirus, RSV
	MX-PCR	Real-time	Adenovirus, human bocavirus
	MX-PCR	Microarray	Influenza A and B, parainfluenza 1–3, RSV A and B, human metapneumovirus A and B, adenovirus A, B, C, and E, rhinovirus A and B, enterovirus A-D, coronavirus Oc43, 229E, NL63, and HKU1
2010	MX-PCR	Real-time	Influenza A, B, and C
	Quantitative MX-PCR	Real-time	RSV, influenza A and B, parainfluenza 1–3, coronaviruses NL63, HKU1, 229E, and OC43; human metapneumovirus, bocavirus
	MX-PCR	Real-time	RSV, human metapneumovirus
	MX-PCR	Real-time	Influenza A and B, RSV, typing of influenza H1N1

(*Continued*)

Tab. 10.3: (*Continued*)

Year of introduction	Technique	Detection procedure	Pathogens targeted
	MX-PCR	Real-time	*M. pneumoniae, C. pneumoniae,* influenza A and B, parainfluenza 1–3, RSV, rhinovirus, enterovirus, adenovirus, human metapneumovirus, coronavirus NL-63, 229E, and OC43
	MX-PCR	Microarray	*M. pneumoniae, C. pneumoniae,* influenza A, B, and C, parainfluenza 1–4, RSV A and B, rhinovirus, adenovirus, human metapneumovirus A and B, bocavirus, coronavirus NL-63, 229E, HKU-1, and OC43
2011	MX-PCR Quantitative PCR	Real-time TaqMan low-density array	Influenza A and B, RSV *M. pneumoniae, C. pneumoniae, H. influenzae, L. pneumophila, S. pneumoniae, S. pyogenes, B. pertussis,* influenza A and B, RSV, parainfluenza 1–3, human metapneumovirus, rhinovirus, enterovirus, parechovirus, adenovirus
	4-tube MX-PCR	Real-time	Influenza A and B, enterovirus, adenvirus, RSV, rhinovirus, coronavirus, human metapneumovirus, parechovirus, parainfluenza 1–4, bocavirus
	MX-PCR	Electrospray ionization mass spectrometry	Influenza A and B, parainfluenza 1–4, RSV, adenoviridae A-F, coronaviridae, human metapneumovirus, bocavirus
	MX-PCR	Real-time	Influenza A and B, RSV, typing influenza H3N2 and H1N1
2012	MX-PCR	ELISA	Influenza A and B, parainfluenza 1-3, rhinovirus, human metapneumovirus, adenovirus, RSV A/B, coronavirus NL-63, 229E, HKU-1 and OC43, bocavirus
	MX-PCR	Microarray	influenza A, B and C, parainfluenza 1-4, RSV A/B, rhinovirus, adenovirus, enterovirus, bocavirus, coronavirus 229E, human metapneumovirus

(*Continued*)

Tab. 10.3: (*Continued*)

Year of introduction	Technique	Detection procedure	Pathogens targeted
	MX-PCR	Real-time	Influenza A and B, RSV A and B
	MX-PCR	Nanoliter real-time	*B. pertussis, B. parapertussis*
	MX-PCR	Nanoliter real-time	RSV A and B
	MX-PCR	Real-time microarray	Adenovirus, bocavirus, coronavirus NL-63, 229E, HKU1, and OC43, enterovirus, human metapneumovirus, influenza A and B, parainfluenza 1-4, rhinovirus, RSV A and B, *B. pertussis, C. pneumoniae, L. pneumophila, M. pneumoniae*
2013	MX-PCR	Dipstick-based molecular test	Influenza A and B
	MX-PCR	Microarray	76 viruses
	MX-PCR	Real-time	Influenza A and B, RSV A and B, human metapneumovirus, parainfluenza 1-4, rhinovirus, enterovirus, coronavirus NL-63, 229E, HKU1, and OC43
	MX-PCR	Electrophoresis	Influenza A and B, parainfluenza 1-3, rhinovirus, human metapneumovirus, adenovirus, coronavirus NL-63, 229E, HKU1, and OC43, RSV A and B, bocavirus
	MX-PCR	ELISA	Influenza A and B, parainfluenza 1-4, RSV, enterovirus, rhinovirus, human metapneumovirus, coronavirus 229E and OC43, reovirus, adenovirus, *B. pertussis, B. parapertussis, C. pneumoniae, L. pneumophila, M. pneumoniae*

Tab. 10.4: Commercially available mono and multiplex nucleic acid amplification based tests for detection of respiratory pathogens.

Manufacturer	Kit	Technique	Detection procedure	Pathogens targeted
Abbott Molecular	PLEX-ID/Flu assay	MX-PCR	Electrospray ionization mass spectrometry	Influenza A (H1N1, H3N2) and B
Affigene	Cp/Mp tracer	MX-PCR	Real-time	*M. pneumoniae and C. pneumoniae*

(*Continued*)

Tab. 10.4: (*Continued*)

Manufacturer	Kit	Technique	Detection procedure	Pathogens targeted
AID GmbH	AID CAP viral assay	MX-PCR	ICT	Influenza A and B, parainfluenza 1–3, RSV A/B, adenovirus
AID GmbH	AID CAP juvenile assay	MX-PCR	ICT	*S. pneumoniae, H. influenzae, B. pertussis, B. parapertussis*
AID GmbH	AID CAP bacterial assay	MX-PCR	ICT	*S. pneumoniae, H. influenzae, M. catarrhalis, C. pneumoniae, M. pneumoniae, L. pneumophila*
AusDiagnostics	Easy-Plex Respiratory Pathogen 12 kit	MX-PCR	Real-time	Influenza A and B, parainfluenza 1-4, bocavirus, human metapneumovirus, RSV, enterovirus, rhinovirus, adenovirus, coronaviruses NL63, OC43, 229E, HKU1
BD	*M. pneumoniae* BDProbeTec ET	SDA	Fluorescence	*M. pneumoniae*
BD	*B. pertussis* BDProbeTec ET	SDA	Fluorescence	*B. pertussis*
BD	*L. pneumophila* BDProbeTec ET	SDA	Fluorescence	*L. pneumophila*
BD	*Chlamydiaceae* BDProbeTec ET	SDA	Fluorescence	*Chlamydiaceae*
BioFire Diagnostics	FilmArray RP	MX-PCR	Melting curve analysis	Influenza A and B, adenovirus, human coronaviruses NL63 and HKU1, human metapneumovirus, parainfluenza 1-4, RSV, rhinovirus/enterovirus
bioMérieux/Argene	Chlamylege	MX-PCR	Hybridization	*C. pneumoniae, M. pneumoniae, Legionella* spp.
bioMérieux/Argene	Influenza A(M) Group & H1N1 2009 r-gene™	MX-PCR	Real-Time	Influenza A and Influenza A H1N1 2009
bioMérieux/Argene	Influenza A/B r-gene™	MX-PCR	Real-Time	Influenza A and Influenza B

(*Continued*)

Tab. 10.4: (*Continued*)

Manufacturer	Kit	Technique	Detection procedure	Pathogens targeted
bioMérieux/Argene	RSV/hMPV r-gene™	MX-PCR	Real-Time	RSV (A, B) and Metapneumovirus (A, B)
bioMérieux/Argene	Rhino&EV/Cc r-gene™	MX-PCR	Real-Time	Rhinovirus/Enterovirus & Control Cell
bioMérieux/Argene	AdV/hBoV r-gene™	MX-PCR	Real-Time	Adenoviruses (A to G) and Bocaviruses (1 to 4)
bioMérieux/Argene	HCoV/PIV r-gene™	MX-PCR	Real-Time	Coronaviruses (229E, OC43, HKU1, NL63) and Parainfluenza 1 to 4
bioMérieux/Argene	Legio. pneumo/Cc r-gene™	MX-PCR	Real-Time	*Legionella pneumophila* & *Control Cell*
bioMérieux/Argene	Chla/Myco pneumo r-gene™	MX-PCR	Real-Time	*C. pneumoniae* and *M. pneumoniae*
bioMérieux/Argene	*Bordetella* R-gene™	MX-PCR	Real-Time	*Bordetella (IS481)*
bioMérieux/Argene	*Bordetella parapertussis r-gene™*	MX-PCR	Real-Time	*Bordetella parapertussis*
Cepheid	ASRRSV	PCR	Real-time	RSV
Cepheid	ASRMPN	PCR	Real-time	*M. pneumoniae*
Cepheid	Xpert Flu A	PCR	Real-time	Influenza A H1N1
Cepheid	ASR Flu A/B	MX-PCR	Real-time	Influenza A and B
Cepheid	ASRBP2	MX-PCR	Real-time	*B. pertussis*, *B. parapertussis*
Fast-track Diagnostics	FTD Respiratory pathogens 33	MX-PCR	Real-time	Influenza A, H1N1, influenza B, influenza C, rhinovirus, coronavirus NL63, 229E, OC43, HKU1, parainfluenza 1-4, human metapneumovirus, bocavirus, *M. pneumoniae*, RSV A/B, adenovirus, enterovirus, parechovirus, *C. pneumoniae*, *S. aureus, S. pneumoniae*, *Haemophilus influenzae* A and B, cytomegalovirus, *Pneumocystis jirovecii*, *Bordetella* spp. (except *Bordetella parapertussis*), *Moraxella catarrhalis*, *Klebsiella pneumoniae*, *Legionella* spp., *Salmonella* spp.

(*Continued*)

Tab. 10.4: (*Continued*)

Manufacturer	Kit	Technique	Detection procedure	Pathogens targeted
Focus Diagnostics	Simplexa Bordetella Universal Direct Kit	Duplex PCR	Real-time	*Bordetella pertussis, Bordetella parapertussis*
Focus Diagnostics	Simplexa Flu A/B & RSV Direct Kit	MX-PCR	Real-time	Influenza A, influenza B, RSV
GenMark	eSensor Respiratory Viral Panel	MX-PCR	eSensor technology	Adenovirus (B/E and C), influenza A and B, human metapneumovirus, parainfluenza 1-3, RSV A and B, rhinovirus
Genomica	Clart PneumoVir	MX-PCR	Microarray	Influenza A, B and C, RSV A and B, adenovirus, parainfluenza 1–4, rhinovirus, coronavirus 229E, human metapneumovirus A and B, bocavirus, enterovirus
Gen-Probe/ Prodesse	ProParaflu+	MX-PCR	Real-time	Parainfluenza 1–3
Gen-Probe/ Prodesse	ProFlu+	MX-PCR	Real-time	Influenza A and B, RSV
Gen-Probe/ Prodesse	ProhMPV+	MX-PCR	Real-time	Human metapneumovirus
Gen-Probe/ Prodesse	ProAdeno+	MX-PCR	Real-time	Adenovirus 1–51
Gen-Probe/ Prodesse	ProFast+	MX-PCR	Real-time	Influenza A – seasonal A/H1, seasonal A/H3, and 2009 H1N1
Gen-Probe/ Prodesse	ProPneumo1+	MX-PCR	Real-time	*M. pneumoniae, C. pneumoniae*
Idaho Technologies Inc.	R.A.P.I.D. influenza	PCR	Real-time	Influenza A
Idaho Technologies Inc.	FilmArray RP	MX-PCR	Microarray	Influenza A H1N1, influenza A H1, influenza A H3, influenza B, RSV, human metapneumovirus, coronavirus NL63, OC43, 229E, HKU1, adenovirus parainfluenzavirus 1–4, bocavirus, rhino/ enterovirus, *B. pertussis, M. pneumoniae, C. pneumoniae*

(*Continued*)

Tab. 10.4: (*Continued*)

Manufacturer	Kit	Technique	Detection procedure	Pathogens targeted
Immunodiagnostik	MutaPLATE *Chlamydia pneumoniae*	PCR	Real-time	*C. pneumoniae*
Luminex Corp.	xTAG RVP	MX-PCR	Luminex technology	Influenza A (H1, H3, H5, non-specific) and B, RSV A/B, parainfluenza 1–4, adenovirus, rhinovirus/enterovirus, five coronaviruses, human metapneumovirus
Meridian Biosciences	illumigene Mycoplasma	LAMP	Turbidity	*M. pneumoniae*
Minerva BioLabs	Venor MP	PCR	Agarose gel electrophoresis and real-time	*M. pneumoniae*
Minerva BioLabs	Onar LP	PCR	Real-time	*L. pneumophila*
Nanogen Inc.	MGB Eclipse Flu A/B RUO	MX-PCR	Real-time	Influenza A and B
Nanogen Inc.	NGEN RVA ASR	MX-PCR	Microarray	Influenza A and B, RSV A/B, parainfluenza 1–3
Nanosphere Inc.	Verigene SP	MX-PCR	Nanoparticle probe technology	Influenza A and B, RSV A/B
PathoFinder	RealAccurate Respiratory RT PCR kit v2.0	Mono- or duplex-PCR (9 individual assays)	Real-time	Influenza A and B, RSV A and B, parainfluenza 1–4, coronaviruses 229E and OC43, rhinovirus/enterovirus, human metapneumovirus, adenovirus
PathoFinder	RespiFinder 19	MX-PCR	Capillary electrophoresis	Influenza A and B, influenza A H5N1, RSV A and B, human Metapneumovirus, *L. pneumophila*, *B. pertussis*, *M. pneumoniae*, *C. pneumoniae*, parainfluenza 1–4, rhinovirus, coronaviruses 229E, NL63 and OC43

(*Continued*)

Tab. 10.4: (*Continued*)

Manufacturer	Kit	Technique	Detection procedure	Pathogens targeted
Patho Finder	RespiFinder 22	MX-PCR	Capillary electrophoresis or microfluidics	Influenza A and B, influenza A H1N1v, RSV A and B, human Metapneumovirus, *L. pneumophila*, *B. pertussis*, *M. pneumoniae*, *C. pneumoniae*, parainfluenza 1–4, rhinovirus/enterovirus, coronaviruses 229E, NL63, OC43 and HKU1, bocavirus
Qiagen	artus Influenza LC RT-PCR	PCR	Real-time	Influenza A and B
Qiagen	artus Bordetella LC PCR kit	MX-PCR	Real-time	*B. pertussis*, *B. parapertussis*, *B. bronchiseptica*
Qiagen	ResPlex II	MX-PCR	Luminex technology	Influenza A and B, RSV A/B, parainfluenza 1–4, human metapneumovirus, rhinovirus, Coxsackie and echovirus
Roche	Influenza A/H1N1	PCR	Real-time	Influenza A H1N1
Seegene Inc.	Seeplex RV15 ACE	MX-PCR	Capillary electrophoresis	Influenza A and B, RSV A/B, parainfluenza 1–4, rhinovirus, three coronaviruses, adenovirus, bocavirus, enterovirus
Seegene Inc.	Seeplex Pneumobacter ACE	MX-PCR	Capillary electrophoresis	*S. pneumoniae*, *H. influenzae*, *M. pneumoniae*, *C. pneumoniae*, *L. pneumophila*, *B. pertussis*
Seegene Inc.	Anyplex II RV16	MX-PCR	Real-time	Influenza A and B, RSV A and B, parainfluenza 1–4, adenovirus, human metapneumovirus, rhinovirus/enterovirus, coronaviruses 229E, NL63, OC43 and HKU1, bocavirus

(*Continued*)

Tab. 10.4: (*Continued*)

Manufacturer	Kit	Technique	Detection procedure	Pathogens targeted
Seegene Inc.	Magicplex	MX-PCR	Real-time	Influenza A and B, RSV A and B, parainfluenza 1–4, adenovirus, human metapneumovirus, coronaviruses 229E, NL63 and OC43, bocavirus, picornavirus, *M. pneumoniae*, *C. pneumoniae*, *B. pertussis*
TIB MolBIOL	*L. pneumophila* LightMix kit	PCR	Real-time	*L. pneumophila*
TIB MolBIOL	Influenza LightMix kit	PCR	Real-time	Influenza A

ICT: immunochromatography; MX-PCR: multiplex PCR; SDA: strand displacement amplification.

conventional and real-time multiplex protocols. However, comparison between mono- and multiplex assays has been performed rarely. Findings and conclusions result frequently in contradictory and conflicting data concerning the sensitivity and specificity of the MX-NAATs compared with the mono NAAT. Increasing the number of targets in one reaction results in loss of sensitivity because the presence of several pairs of primers increases the probability of mispairing resulting in non-specific amplification products and the formation of primer–dimers. Furthermore, enzymes, primers, and salt concentrations as well as cycle protocols required for each target may be different. Results of proficiency panels seem to confirm that conventional multiplex assays may be less sensitive than monoplex assays but as the number of organisms present in clinical specimens of patients is not known, it is impossible to state whether the degree of sensitivity is clinically acceptable.

Similar to conventional single target tests, conventional multiplex reactions are increasingly replaced by real-time multiplex assays because of their greater user friendliness. Real-time MX-PCRs have been applied to two to three agents simultaneously, for example for detection of influenza A virus, influenza B virus, and RSV. The number of agents that can be detected simultaneously in one real-time multiplex reaction tube is restricted by the number of available wavelengths in existing equipment (mostly three at present). However, several reaction tubes can be run in parallel: two or more triplex reactions are often combined in order to increase the number of targets detected. For instance, real-time MX-PCR for the diagnosis of influenza

A virus, influenza B virus, and RSV is performed in a first tube and the four parainfluenza viruses in a second tube. The major drawbacks of this approach are the higher hands-on time required to manipulate all the tubes and a suboptimal cycle protocol for some agents involved in the assay. A compromise must be made considering the optimal temperature cycling requirements and the sensitivity of each component. More investigations are needed to generate protocols with a minimal loss in sensitivity and shorter hands-on time required for manipulation.

The majority of single target and multiplex reactions for the detection of respiratory agents are in-house developed tests. Recently, commercialized kits have become available for the detection of bacterial targets only or bacteria in combination with respiratory viruses either in mono or in multiplex formats (Tab. 10.4). One of those assays, the Pneumoplex (Prodesse, Milwaukee, WI, USA) targeting seven respiratory pathogens was included in a quality control program. Although the limit of detection of this assay was reported to be 5 CFU/ml for *M. pneumoniae*, 0.01 TCID$_{50}$/ml for *C. pneumoniae*, and 10 copies of recombinant DNA for each organism, the test did not perform well in this program. Furthermore, some commercially available multiplex tests are still technically demanding, requiring 3–4 h hands-on time.

The comparative analysis of the limits of detection of the ResPlex I assay (Qiagen, Hilden, Germany) and real-time single PCR assays demonstrated that the MX-PCR assay is 10-fold less sensitive in detecting *M. pneumoniae*. Furthermore, the ResPlex I assay was performed on a number of swab specimens known to be positive by qPCR for three pathogens (*C. pneumoniae*, *M. pneumoniae*, and *S. pneumoniae*) and detected 50, 59, and 81% of the *C. pneumoniae*-, *M. pneumoniae*-, and *S. pneumoniae*-positive samples, respectively.

The Respiratory Multi Well System (M.W.S.) r-gene (bioMerieux/Argene, Varilhes, France) concept allows the detection of numerous pathogens (adenovirus, bocavirus, *Chlamydophila pneumoniae*, coronavirus, enterovirus, human metapneumovirus, influenza viruses A and B, *Legionella pneumophila*, *Mycoplasma pneumoniae*, parainfluenza virus, respiratoy syncytial virus, and rhinovirus) in the same run. This new concept of duplex PCR working on uniform procedures brings flexibility to the multiplex approach. In addition, the diagnostic strategy can be adapted to the season searching for the most likely pathogens as a first approach. In immunocompromised or severely ill patients, pathogens may be searched systematically as add-on diagnostics. The Respiratory M.W.S. r-gene concept includes a cell control that validates the presence of cells in the sample. False-negative results due to a bad sampling can be excluded by this cell control.

Recently, an extraction-independent multiplex RT-qPCR assay for rapid detection of influenza A and B viruses and RSV, the Simplexa Influenza A/B & RSV Direct kit (Focus Diagnostics, Cypress, CA) was brought on the market. Without extraction, nasopharyngeal swab specimens are directly pipetted into the sample wells of the specially-designed "Direct Amplification Disc" for qPCR amplification and target detection.

The latest evolution combines conventional PCR with microarray detection as recently described: NGEN RVA ASR kit (Nanogen Inc., San Diego, CA, USA) and the ResPlex II assay (Qiagen), detecting seven and 12 respiratory viruses or virus groups, respectively (by microarray and Luminex liquid chip hybridization and identification, respectively). Sensitivities of the NGEN and ResPlex II assays were also lower compared with those of monoplex real-time reversed transcriptase PCR assays, most noticeably for RSV and HPIV-3. Although these may be improved by further primer/probe optimization, changes in primer/probe sequences could negatively influence other assays targeted in the multiplexed reaction. Although hands-on times of these tests are only approximately 60 min, turnaround times are 6 h for the ResPlex II and 9 h for the NGEN kit.

Several studies comparing MX-NAATs have been published recently. These studies have shown differences regarding sensitivity for individual viruses while the specificity was usually excellent. In the large European GRACE study, similar results with large differences in amplification efficacy were found when several commercially available tests were compared. However, further studies including an extended number of clinical specimens with low concentrations of the pathogens are required. Furthermore, it appears to be important to develop effective clinical and laboratory algorithms for the optimal use of those assays and to study their impact on patient care in different populations in different settings.

10.4 External quality control

Because only few CE/IVD labeled and/or FDA approved NAATs for detection of respiratory agents are commercially available, laboratories must develop and verify their NAAT results. Quality assurance (QA) programs are strongly required and need to include all aspects of the assay from specimen extraction to detection of amplification products. In particular, the performance of in-house assays is often assessed with samples that fail to resemble clinical specimens, are limited in number, and include too few samples per panel. In addition, in-house testing often does not challenge reproducibility, sensitivity, or turnaround time. For these reasons, there is an urgent need for standardization of methods and reference reagents as well as for complete QA programs including proficiency panels.

To date, only very few studies are available comparing results of different NAATs for the detection of respiratory pathogens in different centers. Only two studies describing multicenter comparisons of the performance of various NAATs for detection of *M. pneumoniae*, *C. pneumoniae*, and *L. pneumophila* in respiratory specimens have been published. Each of these studies revealed significant inter-center discordance of detection rates, using different, or even the same tests, despite the fact that the laboratories participating were experienced in the use of PCR assays. These inconsistent results illustrate the importance of adequate training of personnel and the use of negative and positive controls in the preparative, amplification, and detection phases, shown by the fact that some laboratories reporting false-positive results did not use negative controls in the preparation, amplification, and detection procedures.

With proficiency testing programs, not only performances of laboratories but also characteristics of the NAATs used can be evaluated. Inconsistent findings may be explained by different sensitivity and specificity issues of the tests used. These should be addressed before publishing clinical and epidemiological studies on the role of etiologic agents in LRTIs based on the detection of DNA or RNA in clinical specimens by NAAT. From the above-mentioned multicenter comparisons, it can be concluded that multiplex assays are less sensitive than MX-NAATs with one of the MX-PCRs being a commercially available assay. Analogous results are available from a proficiency-testing panel for the detection of respiratory viruses. With all of the NAATs used, false-positive and false-negative results were generated. A commercial assay evaluated was significantly less sensitive failing to detect the low positive members of the panel; based on these results, the assay is under improvement now. These studies further underline the need for reference reagents and standard operating procedures including quality control assessment of NAATs. Proficiency panels are provided, for instance, by Quality Control for Molecular Diagnostics (QCMD, Glasgow, Scotland, UK).

10.5 The clinical usefulness and implementation of NAATs

The addition of PCR-based methods to the conventional microbial techniques has improved the detection of etiologic agents significantly. Furthermore, it has proven that PCR is not only more rapid than conventional methods but also more sensitive, in etiologic diagnosis of CAP and for the detection of respiratory viruses in LRTI, allowing clinicians to start optimal antibacterial treatment early with rational use of antibiotics and to initiate adequate antiviral therapy when indicated.

The use of single target assays or of MX-NAATs has increased the diagnostic yield in respiratory infections by 30–50%. In combination with traditional bacteriologic techniques to diagnose *S. pneumoniae* infections, more than 50% and in some studies of CAP up to 70% of etiologic agents can be detected. The wider application of multiplex reactions during recent years also resulted in the detection of numerous simultaneous viral infections with widely varying incidences. These differences may result from the variety of diagnostic panels applied. There were no preferential combinations of viruses. Only a few studies found combined infections to be associated with a more severe clinical presentation. The clinical significance of combined infections remains to be further clarified.

Respiratory viruses have also been increasingly recognized as causes of severe LRTIs in immunocompromised hosts. Respiratory infections are more common in solid organ recipients, particularly in lung transplant recipients. Infections are especially dangerous prior to engraftment and during the first 3 months after transplantation, in the setting of graft versus host disease. The origin of infections is community-acquired as well as nosocomial.

Presently, NAATs are more expensive than conventional approaches with the most expensive being the fluorogenic-based real-time detection systems. However, improvements in standardization and automation for sample preparation and technical advances in thermocyclers allowing multiple runs of PCR simultaneously or in a very short time will lead to increased use of amplification methods and cost reductions to rates competitive with conventional methods.

Several studies tended to show cost efficiency of rapid diagnosis of ARI through reduction of antibiotic use and complementary laboratory investigations but most significantly through shorter hospitalization and reduced isolation periods of patients. Despite the probability that improved patient outcome and reduced cost of antimicrobial agents, reduced use of less sensitive and specific tests, and reduced length of hospital stay will outweigh the increased laboratory costs incurred by molecular testing, such savings are difficult to document, and further studies are urgently required.

During epidemics, it may be as important to rule out a particular infection. A considerable saving of diagnostic procedures in ARI is possible by the abolishment of tissue cultures and serologic tests. A closer collaboration between clinicians and the laboratory has a high priority.

10.6 Concluding remarks

With the use of new tools such as NAATs, a better understanding of the etiology and epidemiology of RTIs is possible. However, a number of issues remain to be investigated. The implementation of quantitative tests may shed further light on the relation between virus load and the severity of the disease, produce useful prognostic information, and help in the differentiation between colonization and infection. Since NAATs are more sensitive than tissue culture tests, they may offer more information on the length of the post-infection carrier state and subclinical infections including their involvement in spreading infections. The role of viruses in chronic respiratory diseases such as chronic obstructive pulmonary disease and cystic fibrosis should also be better evaluated. The rapid molecular characterization of the previously unknown SARS-coronavirus within a few weeks after the appearance of the disease and the recent discovery of bocavirus illustrate the potency of NAATs for broadening the knowledge on unknown viruses remaining to be discovered.

In the future, the need for the detection of an ever-expanding number of infectious agents may exceed the capacity of MX-NAATs based on qPCR. The task will be taken over by the next generation of diagnostics, the array technology that could open a wide access to infectious agents.

10.7 Further reading

[1] Bamberger, E.S., Srugo, I. (2008) What is new in pertussis? Eur. J. Pediatr. 167:133–139.
[2] Ieven, M. (2007) Currently used nucleic acid amplification tests for the detection of viruses and atypicals in acute respiratory infections. J. Clin. Virol. 40:259–276.
[3] Mahony, J.B. (2008) Detection of respiratory viruses by molecular methods. Clin. Microbiol. Rev. 21:716–741.
[4] Loens, K., Goossens, H., Ieven, M. (2010) Acute respiratory infection due to *Mycoplasma pneumoniae*: current status of diagnostic methods. Eur. J. Clin. Microbiol. Infect. Dis. 29: 1055–1069.
[5] Loens, K., Mackay, W.G., Scott, C., Goossens, H., Wallace, P., Ieven, M. (2010) A multicenter pilot external quality assessment programme to assess the quality of molecular detection of *Chlamydophila pneumoniae* and *Mycoplasma pneumoniae*. J. Microbiol. Meth. 82:131–135.
[6] Loens, K., Ursi, D., Goossens, H., Ieven, M. (2003) Molecular diagnosis of *Mycoplasma pneumoniae* in respiratory tract infections. J. Clin. Microbiol. 41:4915–4923.
[7] Loens, K., Van Heirstraten, L., Malhotra-Kumar, S., Goossens, H., Ieven, M. (2009) Optimal sampling sites and methods for detection of pathogens possibly causing community-acquired lower respiratory tract infections. J. Clin. Microbiol. 47:21–31.
[8] Loens, K., van Loon, A.M., Coenjaerts, F., van Aarle, Y., Goossens, H., Wallace, P., Claas, E.J., Ieven, M., on behalf of the GRACE study group. (2011) Performance of a multipathogen External Quality Assessment (EQA) panel by different mono- and multiplex nucleic acid amplification tests. J. Clin. Microbiol. doi:10.1128/JCM.00200-11.
[9] Vernet, G., Saha, S., Satzke, C., Burgess, D.H., Alderson, M., Maisonneuve, J.F., Beall, B.W., Steinhoff, M.C., Klugman, K.P. (2011) Laboratory-based diagnosis of pneumococcal pneumonia: state of the art and unmet needs. Clin. Microbiol. Infect. 17:(Suppl 3)1–13.
[10] Woodhead, M., Blasi, F., Ewig, S., Garau, J., Huchon, G., Ieven, M., Ortqvist, A., Schaberg, T., Torres, A., van der Heijden, G., Read, R., Verheij, T.J. (2011) Joint Taskforce of the European Respiratory Society and European Society for Clinical Microbiology and Infectious Diseases. Guidelines for the management of adult lower respiratory tract infections – full version. Clin. Microbiol. Infect. Nov.17:Suppl6:E1–59.
[11] Yang, G., Erdman, D.E., Kodani, M., Kools, J., Bowen, M.D., Fields, B.S. (2011) Comparison of commercial systems for extraction of nucleic acids from DNA/RNA respiratory pathogens. J. Virol. Meth. 171:195–199.

11 Molecular diagnosis of gastrointestinal pathogens

Corinne F.L. Amar

Gastroenteritis is an illness characterized by inflammation of tissue lining any part of the gastroenteric tract including the stomach and small and/or large intestines. Gastroenteritis is a common disease which can affect all individuals and which can have an important impact on health and physical development. Worldwide, nearly one in every five childhood deaths – approximately 1.5 million a year – is due to diarrhea, which kills more children than AIDS, malaria, and measles combined.

There is diverse range of microbiological causes for gastrointestinal illnesses that include pathogenic microorganisms and, in some instances, their toxins. Because of this wide range of biological agents, the type of illness, parts of the gastrointestinal tract affected, the inflammatory processes together with the mechanism of pathogenicity of gastrointestinal pathogens are consequently diverse (Tab. 11.1). The most frequently observed symptoms, pathogenicity mechanisms, and most common routes of transmission for representative enteric pathogens are shown in Tab. 11.1.

Gastroenteritis is one of the most common infectious illnesses affecting humans, and globally, it is estimated as the third and second most common cause of death in adults and children less than 5 years of age, respectively. In industrialized countries, between 20 and 35% of the population is affected by one or more episodes of infectious gastroenteritis per year. Although an occasional cause of mortality, in industrialized countries the disease is mainly a socio-economic burden due to treatment (physician, nurse, in-patient and outpatient care, pharmaceutical and diagnostic cost) and loss of productivity associated with time away from paid employment. During 2004, the province of British Columbia (Canada) had a mean annual cost for gastroenteritis estimated at CAN$ 514.2 million (Euro 831.5 million) which represented a mean annual cost per case of CAN$ 1342.57 (Euro 2170.99). In the 1990s, the mean annual cost, expressed in 1994/1995 prices, associated with gastroenteritis in England was estimated at £743 million, the average cost per case of gastroenteritis presenting to the general practitioner (GP) was £253 and the costs of those not seeing a GP were £34. In the developing world, infectious gastrointestinal diseases are of considerably greater morbidity and mortality burden than in industrialized countries. Although there are more studies estimating the incidence of gastroenteritis in industrialized countries, in resource-limited countries where extreme poverty, poor sanitation, crowded living conditions and access to clean water is less common, diarrheal diseases remain a major cause of impaired growth and development and death, especially in children. It has been estimated by Kosek and colleagues for the World Health Organization, that between 1992 and 2000, there was a median of 3.2 episodes of diarrhea per child under 5 years of age in developing areas and countries. The mortality was estimated as 4.9 children per 1000 per year in these regions because of diarrheal illness in the first 5 years of life. Guerrant and colleagues estimated that in 1990, diarrheal diseases accounted for 12,600 deaths per day in children in Asia, Africa, and Latin America combined and was one of the major causes of childhood mortality in the developing world. Gastroenteritis is not only a particular problem in children, but also in some specific vulnerable groups. For example, gastroenteritis is a common illness in patients with AIDS worldwide, which can become life threatening during the course of their illness. According to Guerrant and colleagues, in 1990, at initial presentation, diarrhea was present in 50–60% of patients with AIDS in the USA, and 95% of patients in Africa and Haiti. In Zaire, infants with HIV-1 infection have an 11-fold increased risk of death from diarrhea. However, since 1998–99, the use of the highly active anti-retroviral therapy (HAART) has enabled a significant increase in CD4 T cell counts in HIV patients, and has helped increase the immune response to gastrointestinal pathogen infections in these patients.

Tab. 11.1: Mechanism of pathogenicity, incubation time, major symptoms, at risk group, and transmission routes of common gastrointestinal pathogens.

Organism	Mechanism of pathogenicity	Incubation time	Major symptoms	At risk group	Main routes of transmission
VIRUSES					
Norovirus Rotavirus A Sapovirus	Attachment and invasion of enterocytes. Lysis of the enterocytes and release of new virions	12–72 h	Diarrhea; projectile vomiting (norovirus); dehydration; asymptomatic carriage	All individuals; children <5 years old (rotavirus and sapovirus)	Person to person; water; food
BACTERIA					
Campylobacter jejuni and *C. coli*	Invasion of epithelial cells of the small intestine. Acute inflammation and ulceration of intestinal mucosa	2–10 days	Diarrhea; abdominal pain; fever	Children and young adults	Food; water and milk; direct contact with infected animals
Verocytotoxic *Escherichia coli* (EHEC or VTEC)	Adhesion to cells and production of toxins; systemic infection	3–8 days	Diarrhea; abdominal pain; bloody diarrhea; hemolytic uremic syndrome (HUS)	All individuals; children <5 years (HUS); >60 years (thrombotic thrombocytopenic purpura)	Food; person to person; contact with infected farm animals

(Continued)

Tab. 11.1: (*Continued*)

Organism	Mechanism of pathogenicity	Incubation time	Major symptoms	At risk group	Main routes of transmission
Clostridium difficile	Production of toxins. Inflammation in the large bowel and necrosis of the colonic brush. Massive cell death	Not known. Time from antibiotic exposure: 1 day to 6 weeks	Profuse diarrhea, pseudomembranous colitis	All individuals under therapy with broad spectrum antibiotics and long stay at hospital	Person to person; own microflora
Clostridium perfringens	Production of toxin. Damage of the intestinal mucosa and of enterocytes	8–24 h	Abdominal cramps; nausea and vomiting (less commonly); bloody diarrhea, shock, toxemia and death (*C. perfringens* type C)	All individuals; the elderly and individuals under antimicrobial therapy and long stay at hospital	Food; person to person and own microflora (antimicrobial treatment associated diarrhea)
Salmonella enterica	Invasion of enterocytes and of M cells; systemic infection (*S. Typhimurium*); bacteremia (*S. Typhi, S. Paratyphi*)	12–48 h	Fever and chills; diarrhea; nausea and vomiting; abdominal cramps	All individuals	Food
PROTOZOAE					
Cryptosporidium	Not known. Intestinal malabsorption and damage to the enterocytes	1–3 weeks	Diarrhea; abdominal cramps; bloating; pronounced fatigue; weight loss; malabsorption of fat-soluble vitamins; asymptomatic carriage	Children <1–5 years; individuals in contact with young children	Water; food; person to person; contact with infected animals
Gialdia					

Because the symptoms of gastroenteritis are not usually specific for different etiological agents, accurate laboratory-based diagnostics provides important information to allow appropriate interventions that include antimicrobial-treatment, control of secondary spread and withdrawal of sources of infection to prevent additional illnesses. This chapter will consider the diagnosis of gastroenteritis using molecular methods.

11.1 Clinical manifestations

Symptoms of gastroenteritis depend on the degree of inflammation and usually include vomiting and/or diarrhea, the latter comprising the frequent passages of watery or semi-formed stools with or without mucus or blood although some infections will be asymptomatic. A case-control study of gastroenteritis performed in 2007 in England revealed that more than 40% of the asymptomatic population carries at least one enteric pathogen. Additional symptoms of gastroenteritis include fatigue, fever, flatulence, abdominal pain, nausea, malaise, and/or general discomfort. Dehydration is the most frequent sequela, especially in infants with frequent and abundant diarrhea or vomiting.

There are different forms of the disease caused by infectious agents. Ulcerative colitis, for example, is an ulceration of the large intestine or colon. Pseudomembranous colitis is an inflammation of the colon showing raised plaques or pseudomembranes comprising areas of dead epithelial and blood cells. Necrotizing enteritis affects the small intestine, the jejunum, and the ileum, and presents with necrosis and perforation of the lining of the intestine.

The symptoms of gastroenteritis are usually not specific for different etiological agents. Severity and duration may vary influenced by the agents themselves, the initial level of exposure, and interactions with other pathogens, together with the patient's age, underlying illnesses, diet, existing enteric microflora, and immune status. The most frequent symptoms are shown in Tab. 11.1.

11.2 Preanalytics

The most common method for investigating gastrointestinal infections is by the laboratory-based examination of feces for gastrointestinal pathogens or their toxins. The examination of feces has the advantage that the specimen is generally:
- readily available,
- collected, often by the patient themselves, without invasive procedures or additional adverse effects,
- collected and transported without specialist transported media,
- unlikely to present an undue hazard to those transporting the specimen to the laboratory.

There are alternatives to analysis of feces, for example, by examination of material collected during invasive diagnostic procedure (including the collection of biopsy material). However, because of the methodological differences in the analytical techniques used (including histological analysis) and the more limited use, these are not discussed here.

"Normal" feces comprises up to 75% water and 25% solid matter. Approximately 30% of the solid matter consists of bacteria, 30% of food components including 10–20% fats, 10–20% inorganic substances, and 2–3% proteins. The remainder is made up of host material including cells from the gastrointestinal tract, bile, enteric secretions, leukocytes, bilirubin, and products resulting from the action of the microflora. Fecal specimens can be highly heterogeneous and the microbial composition of feces is extremely variable depending on the individual, diet, age, and even time of the day of collection. The microflora of feces provides an extremely dynamic environment, which can have marked effects on the integrity of gastrointestinal pathogens or microbial toxins prior to examination. Furthermore, this specimen is not directly obtained from the affected anatomical site (i.e. the gastrointestinal mucosa), the presence and concentration of potential pathogens in the feces do not necessarily reflect that at the affected site.

The preservation of gastrointestinal pathogens in the feces and the subsequent recovery of their nucleic acid can be affected by condition of storage before processing. Remarkably, limited investigation has been performed on the degradation of gastrointestinal pathogens and their toxins in feces under different conditions of storage. However, it is recognized that *Campylobacter* and the parasites *Dientamoeba fragilis* and *Giardia duodenalis* will rapidly die and their DNA become degraded after storage in whole feces.

The principles of preservation of gastrointestinal pathogens in fecal specimens include:

- Enteric pathogens do not replicate in feces during transport at room temperature, since this is not their optimal living environment. Some protozoa, how-ever, undergo maturation in feces before they become infectious (e.g. *Cyclospora cayatenensis*).
- Although often present in high levels during acute infection, enteric pathogens can be shed intermittently and in low numbers especially during asymptomatic or subclinical infections or in the convalescent phase.
- Bacteria, viruses, and some protozoa are generally susceptible to subsequent freeze–thawing treatment.
- Conventional diagnostic procedures including production of environments, which permit only the growth of specific bacterial pathogens, might hamper the recognition of co-infections.
- The action of proteases and nucleases in feces may degrade targets essential for diagnostic procedures.

Consequently, fecal samples should be examined as soon as possible after collection and ideally stored at 2–4°C. However, this is usually not possible from the point of collection to the laboratory where transport at ambient temperatures is most usually practiced. Preservation agents (e.g. formalin, potassium dichromate, or antimicrobial solutions) are sometimes added, particularly by parasitologists or virologists; however, these are not suitable for some diagnostic procedures and are likely to reduce the sensitivity of PCR-based methods.

11.3 Analytics

Conventional approaches for the detection of gastrointestinal pathogens include the recognition of the agents themselves by light or electron microscopy; the detection of products specific to their metabolism using immuno- or toxicity assays, or by growing the agents using *in vitro* culture and the recognition of specific phenotypic (or genotypic) characters (Tab. 11.2). Because of the variety of the infectious agents, a range of techniques is required for their detection and hence no single approach will provide an optimal diagnostic strategy. There is therefore a requirement for staff to be skilled in the application and interpretation of multiple techniques with consequential procurement and maintenance of varied equipment. These conventional approaches are generally time consuming, labor intensive, costly and, for some agents, may be unable to distinguish between pathogenic of non-pathogenic variants of the same species (Tab. 11.2).

Tab. 11.2: Methods used in conventional approach for the diagnosis of infectious gastrointestinal diseases.

Methods	Application	Comments
Electron and light microscopy	Visualization of microorganisms; identification from morphology	Slides preparation and staining (Gram, acid fast, immunofluorescence heavy metal staining)
		Requires expensive equipment and dedicated laboratory space
		Specialist training and experience
		Results can be subjective
		Time consuming and of low sensitivity
		New pathogens can be discovered (viruses; parasites)
In vitro culture	Enrichment of bacteria on liquid media; purification of isolates on solid media	Do not apply for non-culturable organisms
		Need a variety of selective media suitable for the growth of the specific bacterial groups
		Need for special equipment to provide suitable temperature and gaseous conditions permitting bacterial growth
		Equipment can be expensive
		Time consuming for "slow" growing bacteria
Phenotypic tests	Identification by detection of end products of metabolism; characterization of enzymatic activities	Need a range of specialized media
		Relies on the purity of the isolate and on growth requirements
		Bacterial metabolism is not always specific since some variants do not express typical phenotypes
		Not always possible to distinguish pathogenic and non-pathogenic members of the same species

(Continued)

Tab. 11.2: (*Continued*)

Methods	Application	Comments
Immunoassays	Detection of bacterial toxins; detection of organism specific antigens	Labor intensive for antigen preparation Production of anti-sera Possible cross-antigenic reactions resulting in false-positive results Low sensitivity resulting in false-negative results Bacterial toxins can be detected in the absence, or presence of the secreting bacteria in low numbers
Cytotoxicity tests	Biological activity of virulence factor directly observed on mammalian tissue cultured *in vitro*	Need to maintain cell lines Relies on the purity of the isolate and on growth requirements The effect observed can be subjective and non-specific The observation is performed under light microscopy and the microscopist needs training and experience It is simple to perform and results can be obtained rapidly New pathogens can be discovered

In contrast, molecular approaches (PCR-based techniques to amplify and detect specific nucleic acid sequences) use generic techniques, which can be substituted for the conventional methods outlined above. These molecular approaches are rapid, sensitive, specific and, albeit that interpretation may differ from conventional techniques, provide unambiguous results (Tab. 11.3). Because of their generic nature, molecular techniques can better support the detection of co-infections which can be frequent in gastrointestinal infections in some patient groups, and are more amenable to the application of automation. PCR-based assays applied to nucleic acid recovered from feces, however, are not without potential disadvantages, most importantly due to spurious nucleic acid cross-contamination from sources other than the sample itself, and co-extraction of substances inhibitory to the PCR. These problems will be discussed later in this section.

The initial stages of sample processing before PCR include the extraction and purification of the enteric pathogens nucleic acid. The purification of good quality nucleic acid containing sufficient copies of the target pathogen sequences and intact enough to allow subsequent amplification is essential to obtain optimal PCR results. For detection of DNA targets, these extracts may be suitable for direct analysis by PCR. However, for RNA (viral) targets, reverse transcription (RT) of the nucleic acid extract will be necessary for transcription into complementary DNA (cDNA) which can be subjected to PCR.

The first stage of extraction consists of disruption of enteric pathogens present and the release of their nucleic acid. Various techniques can be used depending on the robustness of the organisms: viruses and Gram-negative bacteria being the least robust and Gram-positive bacteria and protozoan cysts/oocysts the most. Because gastroenteritis can be caused by a diversity of enteric pathogens, the technique or techniques used should allow the disruption of as wide a range of microorganisms

Tab. 11.3: Methods used in the molecular approach for the diagnosis of infectious gastrointestinal diseases.

Methods	Application	Comments
Nucleic acid extraction	Disruption of enteric pathogens and purification of the released nucleic acid	The process is generic, can be automated and allows a high throughput Disruption technique has to suit a variety of microorganisms Co-extraction from feces of inhibitors of the PCR and further purification of the nucleic acid might be required Extracted RNA is not stable and needs to be converted into cDNA by random priming RT
PCR	Specific DNA or cDNA sequence is amplified	Generic, rapid, sensitive, and very specific. Can be automated and therefore support the recognition of co-infections Can be used to detect virulence factors, identify antimicrobial resistance or other target characteristics such as toxin production Prior knowledge of the target is needed Because of the high sensitivity, DNA contamination from other sources can be a problem mRNA amplification for detection of viable microorganisms; does not detect pathogen by-products such as toxins
Gel-based or conventional PCR	Amplified product is visualized and identified by gel electrophoresis	Stringent conditions are applicable for non-subjective results Nested version might be required to increase sensitivity, especially for the detection of enteric pathogens present in low copy numbers
qPCR	Amplification is observed in real time	High throughput Increased sensitivity and specificity compared with gel-based systems Closed tube system means less risk of cross-contamination Gel electrophoresis is no longer needed Quantification of target DNA possible which can differentiate between asymptomatic from symptomatic infections

as possible. The use of two disruption methods in parallel is proposed by the author. The proposed method for extracting DNA or RNA from viruses is based on the use of buffered chaotropic solutions such as guanidinium thiocyanate, which not only inhibits enzymatic activity, including that of nucleases, but also can lyse viral particles. The second proposed method for the disruption of bacteria and protozoans relies on the mechanical disruption of feces with zirconia or glass beads in buffered solution of guanidinium thiocyanate, using machines such as the MagNAlyser®

(Roche) or the FastPrep homogenizer (Thermo Scientific) (Fig. 11.1). Following nucleic acid release, the technique described by Boom and colleagues in 1990 has been successfully applied by the authors and colleagues since 2003 on more than 6000 fecal specimens. This technique showed to be an appropriate method to purify nucleic acid from a wide range of microorganisms from feces. It relies on the capture of DNA and RNA onto silica particles (diatomaceous earth) in the presence of a buffered solution of guanidinium thiocyanate and the subsequent washing of the silica–nucleic acid complex with ethanol. Following suspension of the complex in acetone then drying, the nucleic acid is eluted into water (Fig. 11.1). The same principle of extraction to the method described by Boom and colleagues is used with some commercial systems kits using columns (QIAamp DNA Stool Mini kit, Qiagen) and in automated extraction systems such as the Maxwell®16 (Promega), the BioRobot Universal (Qiagen), the MagNAPure® (Roche) or the Xtractor® (Corbett-Qiagen), which use paramagnetic silica particles or silica membranes. These systems are less labor intensive and improve workflow and turnaround time but may be more expensive.

Problems with recovering nucleic acid from feces include the presence of substances, which can hamper the disruption stage, and the co-extraction of moieties, which inhibit PCR and lead to false-negative results. These PCR inhibitors include by-products of hemoglobin degradation, bile salts, pectins, phenolic compounds, glycogen, formalin, fats, carbohydrates, complex polysaccharides and other macromolecules. Because enteric pathogens or their toxins can alter the absorption and digestion processes in the gastrointestinal tract, these inhibitory substances may be at increased concentrations in feces collected during gastroenteritis to those of normal stools. Various methods have been reported to remove those inhibitors or to prevent their action on the DNA polymerase during amplification. Diluting the DNA extract, and therefore potential inhibitors, can be efficient but there is a risk of reducing the target concentration to below the sensitivity of the assay. The treatment of fecal DNA extracts with chelex resins or with polyvinyl pyrrolidone has been shown to remove or reduce the amount of inhibitors without the loss of target DNA.

Target DNA is generally stable when stored at –80°C to 4°C. However, the authors have experienced spontaneous degradation of giardial and clostridial DNA, even when kept frozen at –20°C or –80°C. Purified RNA, however, is less stable; therefore, for the detection of gastrointestinal RNA viruses, RNA should be converted into cDNA by reverse transcriptase as soon as possible. RT by random priming will allow the transcription of cDNA from all RNA, which then can be used in the detection of a wide range of viruses. This method is therefore advisable in the investigation of enteric pathogens in feces during gastroenteritis. RT by specific priming, such as those used in combination with PCR assays, will limit the range of viruses detected, but may be of greater sensitivity and suitable for detection of restricted targets which are well characterized and show limited or no variation at primer binding sites.

Fecal specimens

Tube with zirconia beads: add 900 µl L6 buffer(a), 20 µl isoamylalcohol (b), and 200 µl of feces

CELLS

Shake at maximum speed for 1 minute

MagNAlyser (Roche Diagnostics)

DISRUPTION

Centrifuge

Prepare 2 ml of a 20 % fecal suspension in salt solution medium. Internal control can be added at this stage

Transfer the supernatant into a 1.5 ml clean tube and add 100 µl silica particles. The internal control can be added at this stage

CAPTURE

200 µl of fecal suspension; 1 ml L6 buffer (a); 20 µl silica particles

Shake gently at room temperature for at least 10 minutes

PURIFICATION

Clean the silica-nucleic acids complex twice with L2 buffer (a) and twice with a solution of 80 % ethanol and once with acetone

Dry the silica for 5–15 minutes

56 °

ELUTION

Elute in sterile, distilled, nuclease-free water

Heat for 5 minutes to help elution

56 °

Elute in sterile distilled water and RNasin

Centrifuge and keep supernatant (nucleic acid)

Perform reverse transcription

PCR

PCR

(a): L6 and L2 are chaotropic buffered solutions of guanidinium thiocyanate
(b): isoanylalcohol is used as an antifoam solution

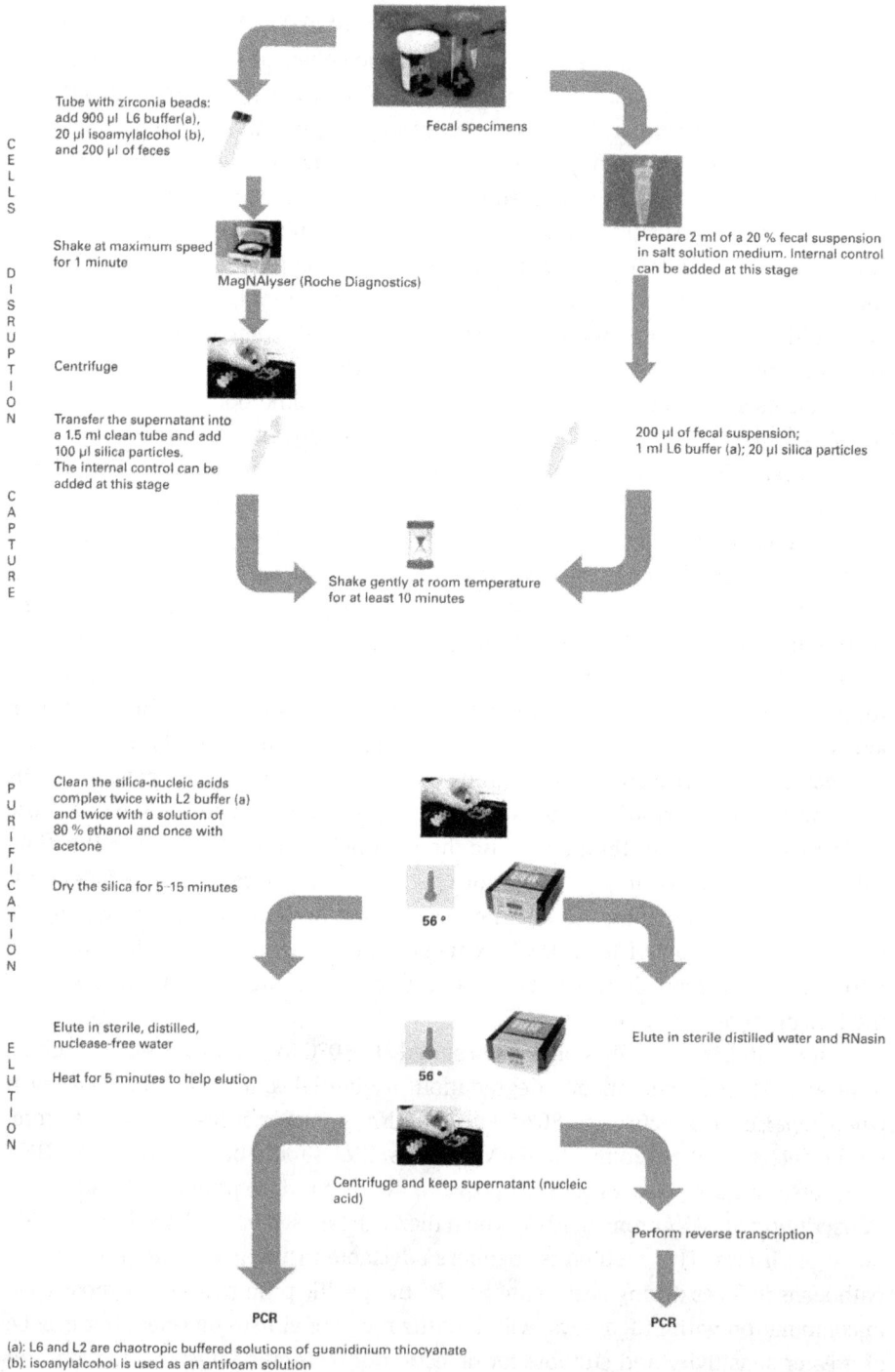

Fig. 11.1: Nucleic acid extraction of enteric pathogens from fecal specimens.

There are two main PCR approaches, which rely on different systems for detection of amplified products: gel-based PCR and real time PCR. In conventional or gel-based systems, amplified target DNA is detected and sized by gel electrophoresis after the PCR reaction has been completed. When used for the detection of gastrointestinal pathogens from feces, highly sensitive gel-based PCR systems, such as nested-PCRs, may need to be developed especially when investigating pathogens that can be present in low copy number such as *Giardia* or *Cryptosporidium*. Because the fecal microflora is very varied and gel-based systems detection relies on sizing of the amplified fragment(s) by gel electrophoresis, stringent conditions are therefore required for optimal specificity, such as the use of high annealing temperature and low magnesium concentration. qPCR is a closed tube system which combines amplification and probe hybridization. Probes labeled with fluorophores are designed to anneal between the primer binding sites and fluorescent signals are generated, recorded, and visualized during amplification. qPCRs have some advantages over gel-based assays, which make them an ideal tool in the investigation of gastrointestinal pathogens directly from feces specimens since:

- Increased sensitivity provides higher chance of pathogen detection, even when the target is present in feces in low number and by-passes the need for nested assays.
- The single closed tube system decreases the possibility of cross-contamination.
- Higher specificity is provided by the primer and the probe sequences and by the detection of the amplification product in real-time, thus providing unambiguous results.
- Quantification of target may allow distinction between symptomatic and asymptomatic carriage of the pathogen (see Postanalytic section).
- Decreased assay time and easier application of automation allow higher throughput and therefore the investigation of a wider range of pathogens.

For the diagnostic laboratory, it is a necessity to have sufficient confidence in the results to provide optimal strategies to be able to monitor the performance of each procedure, and, if possible, each individual assay. One of the main concerns for the application of molecular diagnostics is the control of cross-contamination by exogenous material foreign to the primary sample. This is of particular concern when processing fecal specimens because of the low concentration of target as compared to non-target nucleic acid, most of which will be derived from the microflora. Cross-contamination can arise from a variety of sources including aerosols generated during assay preparation or from previously amplified PCR product. A nanoliter of a typical PCR product contains up to 10^8 target molecules, therefore a dedicated clean DNA-free room for reagents, consumables, and the preparation of PCR reagents mixtures and a unidirectional process to avoid back flow of PCR product is essential. A simple way to monitor cross-contamination is to introduce in each batch of extraction and

PCR one or more non-template controls (NTC), usually made up of all reagents except that water substitutes for the specimen or DNA extract. Any amplification detected in these assays indicates that cross-contamination has occurred and the possibility of false-positive results. The inhibition of target-cells lysis and the co-extraction of inhibitors of the PCR has been mentioned earlier in this chapter and monitoring these is essential to detect false-negative results, that is, failure to detect the target when present. This can be monitored by the use of viral particles, phages, genetically modified bacteria or other nucleic acid sequences which do not occur in any primary sample, and are introduced, either into the fecal specimen (process control), or into each PCR reaction (internal positive control). The failure of a PCR can also be due to human error or to the use of low quality reagents. To monitor this, a positive control made up of the target DNA is introduced in each batch of PCR test.

There is a range of satisfactory gel-based as well as real time PCRs described for gastrointestinal pathogens and some of these are "recommended" through the author's direct experience (Tab. 11.4). However, it is important to understand that it is essential to establish sensitivity and specificity data for each individual assay.

For the detection of gastrointestinal pathogens, laboratory-developed assays are commonly used. For the detection of norovirus RNA, commercial assays have been introduced recently. A summary of both of the commercial assays currently frequently used in Europe is provided in Tab. 11.5.

11.4 Postanalytics

11.4.1 Clinical sensitivity and diagnostic specificity

PCR results obtained from clinical specimens should be interpreted according to:
- the clinical sensitivity or true positive value (the proportion of individuals with the illness and a positive result, i.e. detection of the organism),
- the diagnostic specificity or true negative value (the proportion of individuals without the illness and a negative result, i.e. no detection of the organism) of the assay.

These values can be used to determine the predictive values of a diagnostic assay, that is, an evaluation of how efficient the assay is to diagnose a condition. The clinical sensitivity and the diagnostic specificity are not to be confused with the analytical sensitivity (minimum target copy number to be detected) and specificity (the right target detected) values of the PCR assay, which should both be evaluated, if possible, using quantified target.

Tab. 11.4: Selection of PCRs for the detection of gastrointestinal pathogens, based on the DNA polymerase 5′-exonuclease activity (TaqMan).

Organism	Gene/locus amplified	Encoded protein	F R P	Forward primer sequence Reverse primer sequence Probe sequence	Length of amplification product (bp)	PCR format	Reference
VIRUSES							
Norovirus Genogroup I	ORF1–ORF2 junction	RNA-dependent RNA polymerase	F R **P**	5′-CGYTGGATGCGNTTYCATGA-3′ 5′-CTTAGACGCCATCATCATTYAC-3′ **5′-AGATYGCGRTCYCCTGTCCA-3′**	85	Duplex Fast, exonuclease TaqMan assay	Kageyama et al. (2003) J. Clin. Microbiol. 41:1548–1557
Norovirus Genogroup II			F R **P**	5′-CARGARBCNATGTTYAGRTGGATGAG-3′ 5′- TCGACGCCATCTTCATTCACA-3′ **5′-TGGGAGGGCGATCGCAATCT-3′**	95		
Rotavirus A	VP6	Rotavirus inner capsid protein VP6	F R **P**	5′-GACGGVGCRACTACATGGT-3′ 5′-GTCCAATTCATNCCTGGTGG-3′ **5′-CCACCRAAYATGACRCCAGCNGTARMWG CATTATTTCC-3′**	379	Standard, exonuclease TaqMan assay	Iturriza et al. (2002) J. Virol. Methods 105:99–103
Sapovirus	ORF1 junction	Sapovirus polymerase-capsid junction	F F F	5′-GAYCASGCTCTCGCYACCTAC-3′ 5′-TTGGCCCTCGCCACCTAC-3′ 5′-TTTGAACAAGCTGTGGCATGCTAC-3′	103 and 107	Multiplex Fast, exonuclease TaqMan assay	Oka et al. (2006) J. Med. Virol. 78:1347–1353

(Continued)

Tab. 11.4: (Continued)

Organism	Gene/locus amplified	Encoded protein		Forward primer sequence / Reverse primer sequence / Probe sequence	Length of amplification product (bp)	PCR format	Reference
			R	5'-CCCTCCATYTCAAACACTA-3'			
			P	5'-CCRCCTATRAACCA-3' MGB			
			P	5'-TGCCACCAATGTACCA-3' MGB			
Adenovirus type 40 and 41	Fiber	Long fiber gene	F	5'-CACTTAATGCTGACACGGGC-3'	152	Fast, exonuclease TaqMan assay	Tiemessen et al. (1996) J. Virol. Methods 59:73–82
			R	5'-ACTGGATAGAGCTAGCGGGC-3'			
			P	5'-TGCACCTCTTGGACTAGT-3' MGB			
BACTERIA							
Campylobacter jejuni	mapA	Membrane-associated protein A	F	5'-CTGGTGGTTTTGAAGCAAAGATT-3'	95	Duplex Fast, exonuclease TaqMan assay	Best et al. (2003) FEMS Microbiol. Lett. 229:237–241
			R	5'-CAATACCAGTGTCTAAAGTGCGTTTAT-3'			
			P	5'-TTGAATTCCAACATCGCTAATGTATAAAAG CCCTTT-3'			
Campylobacter coli	ceueE	Enterochelin uptake protein	F	5'-AAGCTCTTATTGTTCTAACCAATTCTAACA-3'	103		
			R	5'-TCATCCACAGCATTGATTCCTAA-3'			
			P	5'-TTGGACCTCAATCTCGCTTTGGAATCATT-3'			
Salmonella enterica	iroB	C-Glycosyltransferase	F	5'-CCAAGAGATCTGGCGTGGATAG-3'	91	Fast, exonuclease TaqMan assay	Murphy et al. (2006) Int. J. Food Microbiol. 120:110–119
			R	5'-GGACGTATTGCATGGAGATAACC-3'			
			P	5'-ACCGCCCAGCATGAGCATACTGC-3'			
EnteroAggregative E. coli	Aa	Anti-aggregation transporter	F	5'-GGGCAGTATATAAACAACAATCAATGG-3'	82	Fast, exonuclease TaqMan assay	Amar et al. (2005) Diagn. Mol. Pathol. 14:90–96
			R	5'-GTAGTTGTTCCTCCTCACTAAGCATTTCAAT-3'			
			P	5'-TCTCATCTATTACAGACACAGCC-3' MGB			

(Continued)

Tab. 11.4: *(Continued)*

Organism	Gene/locus amplified	Encoded protein		Forward primer sequence / Reverse primer sequence / Probe sequence	Length of amplification product (bp)	PCR format	Reference
Verocytotoxic E. coli and Shigella dysenteriae	stx1 (or vt1)	Shiga-toxin 1 subunits A/B	F	5'-GGATAATTTGTTTGCAGTTGATGTC-3'	107	DuplexFast, exonuclease TaqMan assay	Moeller Nielsen et al. (2003) J. Clin. Microbiol. 41:2884–2893
			R	5'-CAAATCCTGTCACATATAAATTATTTCGT-3'			
			P	5'-CCGTAGATTATTAAACCGCCCTTCCTCTGGA-3'			
Verocytotoxic E. coli	stx2 (or vt2)	Shiga-toxin 2 subunits A/B	F	5'-GGGCAGTTATTTTGCTGTGGA-3'	131		
			R	5'-GAAAGTATTTGTTGCCGTATTAACGA-3'			
			P	5'-ATGTCTATCAGGCGCGTTTTGACCATCTT-3'			
PROTOZOAE							
Giardia spp.	ef1a	Elongation factor 1-alpha	F	5'-GTACGAGGGCCCGTGYC-3'	81	Fast, exonuclease TaqMan assay	Amar et al. (2007) Eur. J. Clin. Microbiol. Infect. Dis. 26:311–323
			R	5'-GGGAGRCGRAGRGGCTTGTC-3'			
			P	5'-ATCGACGCGATCGAC-3' MGB			
Cryptosporidium hominis, parvum, meleagridis	Cowp	Cryptosporidium oocyst wall protein	F	5'-GTTCAATCAGACACAGCTCC-3'	107	DuplexFast, exonuclease TaqMan assay	Amar et al. (2007), unpublished
			R	5'-CCAGAAGGACAAACGGTATC-3'			
			P	5'-CCTAATCCAGAATGTCCTCCAGG-3'			
Cryptosporidium felis			F	5'-CAGATACTGCTCCACCAAAC-3'	135		
			R	5'-GATACTGCACGCATCTATTC-3'			
			P	5'-CAGAATGTCCTCCAGGAACAATAC-3'			

Tab. 11.5: Comparison of both of the commercially available assays currently frequently used for the qualitative detection of norovirus RNA in Europe.

Characteristics	Manufacturer and details				
	Altona Diagnostics	AnDiaTec	Cepheid	Immundiagnostik AG	R-Biopharm AG
Kit name	RealStar Norovirus RT-PCR Kit	Norovirus RT-PCR Kit	Smart Norovirus Kit	MutaREX Norovirus real time RT-PCR	RIDAGENE Norovirus I and II
Target sequence	NA	ORF1/ORF2 junction	NA	ORF1/ORF2 junction	ORF1/ORF2 junction
Amplification method	RT qPCR	RT qPCR	RT qPCR	RT qPCR	RT qPCR
Detection method	Fluorescence	Fluorescence	Fluorescence	Fluorescence	Fluorescence
Internal control	Heterologous	Heterologous	Heterologous	Heterologous	Heterologous

NA: data not available; RT: reverse transcription.

The predictive values of a test used to diagnose a condition can only be defined when this condition has been determined by a gold standard method. The gold standard method for diagnosing gastroenteritis consists of the detection of pathogens or the effect of their toxins at the site of infection, that is, the intestinal tract. Because these methods involve bowel biopsy they are usually not performed. Therefore, the predictive values of a non-invasive test for the diagnosis of gastroenteritis, such as a PCR assay, are difficult to establish. However, it should be understood that the clinical sensitivity, or true positive value, of PCRs used in the diagnosis of gastroenteritis are expected to be lower than the analytical sensitivity of the assays. The main factor affecting this sensitivity is the co-extraction of PCR inhibitors and nucleic acid as outlined earlier in this chapter. Because of the heterogeneous nature of feces, the amount of inhibitors co-extracted, and therefore the potential of inhibition, can be variable between specimens but also between extracts recovered from the same specimen. In addition, different PCR assays are not equally susceptible to the same inhibitory substances. The means to overcome the problem of inhibition has been mentioned earlier in the "Analytic section".

Another important factor affecting clinical sensitivity is low target density. Pathogens can be shed intermittently, at low levels, or can form clumps, such as *Cryptosporidium* oocysts. In these situations, the chance of the organism being processed, and therefore being further detected by PCR, can vary. Performing triplicate assays per sample has been shown by the author to be a successful strategy to detect giardial cysts or *Cryptosporidium* oocysts when present in low number (less than 600 units/ml of feces). The heterogeneity of feces and the co-extraction of PCR inhibitors are problems that can be overcome but should not be ignored as part of the overall estimation of uncertainty when interpreting PCR results using DNA recovered from feces.

11.4.2 Interpretation of results

Before interpreting PCR results, it should be clear that a PCR assay is used to amplify a specific DNA or cDNA fragment for which this assay has been designed to detect, and only that fragment. It can be that the target selected is a fragment of a gene coding for a specific protein such as an enzyme or a toxin. Alternatively, it can be a portion of the chromosome specific to an organism. However, because of the specificity of the sequence target, it is essential that each assay is periodically validated (at least *in silico*) against subsequently available target nucleic acid sequences and any genetic variation investigated. Therefore, PCR will detect genetic sequences of target micro-organisms, and not their expressed and biologically active products such as toxins.

As mentioned earlier, stool is not directly obtained from the affected anatomical site and the physiological status of enteric pathogens in stool is different from that at the intestinal site of infection. Therefore, microorganisms which have not survived or which are too damaged to multiply are not recovered by conventional culture but their nucleic acid might be detected by PCR. In addition, PCR allows the investigation of the presence of a wide range of microorganisms. Therefore, PCR results will give more information and may be more complex to interpret than those obtained by conventional microbiology. Recent data generated by PCR showed that more than one pathogen (sometimes more than five) can simultaneously occur in feces, especially in children, and their contacts such as parents and grandparents. A large-scale case-control study in England and Wales, using PCR, has shown that in the community, more than 40% of the asymptomatic population has gastrointestinal pathogens detected in the feces with 16%, 14% and 2% shedding norovirus, rotavirus and *Giardia*, respectively. However, as for conventional microbiology, the interpretation of PCR results for gastrointestinal infections should be made in conjunction with clinical features and severity and characteristics of the illness including time of onset, and accompanying risk factors (e.g. recent travel, farm visit, contact with children or animals, recreational water use, institution catering, restaurant use or the diagnosis from other related cases). If the patient is symptomatic, interpretation should also consider the likelihood that the pathogen or pathogens detected are the cause of the illness.

qPCR has the advantage over conventional or gel-based PCR methods that the relative amount of template DNA, and therefore the pathogen's load, can be quantified. This feature can be useful in distinguishing between symptomatic or asymptomatic infections. For example, norovirus particles can be shed up to 28 days after the onset but the shedding is reduced during the convalescent phase of illness. Using qPCR results from a large scale case-control study, Philips and colleagues in 2009 have determined the PCR threshold point which estimated levels of virus above which, results were most likely to be the cause of illness, and below an indication of asymptomatic carriage. Another example is *Salmonella*, which occurs in feces at high levels during the acute phase and can be shed up to 6 months postinfection. Although there is at present limited data to interpret the pathogen load, it is generally considerably lower in asymptomatic individuals. Further data are needed for the interpretation of relative pathogen load for other agents.

Gastrointestinal infections are common and can be a serious health threat and a significant cause of death, especially in children living in the developing world. It is also a considerable worldwide economic burden. The impact of molecular biology and the development of molecular techniques have allowed a huge advance in the knowledge of microbiology. We have shown in this chapter that the use of PCR-based techniques allows the investigation of a much broader range of pathogens from feces than is usually available through the application of conventional microbiological techniques. In addition, we have shown that PCR is a powerful means to improve the diagnosis of gastrointestinal infectious illnesses in terms of higher sensitivity, specificity, and sample throughput. PCR technology has the potential to increase our understanding and therefore the control of gastrointestinal infectious diseases. Furthermore, as nucleic acid amplification techniques become both simpler and cheaper, their impact is likely to play an increasingly crucial role in research and diagnosis in both the industrialized as well as the developing world.

11.5 Further reading

[1] Amar, C.F.L., East, C.L., Gray, J., Iturriza-Gomara, M., Maclure, E.A., McLauchlin, J. (2007) Detection by PCR of eight groups of enteric pathogens in 4627 faecal samples: re-examination of the English case-control Infectious Intestinal Disease Study (1993–1996). Eur. J. Clin. Microbiol. Infect. Dis. 26:311–323.

[2] Amar, C.F.L., East, C., Maclure, E., McLauchlin, J., Jenkins, C., Duncanson, P., Wareing, D.R.A. (2004) Blinded application of microscopy, bacteriological culture, immunoassays and PCR to detect gastrointestinal pathogens from faecal samples of patients with community-acquired diarrhoea. Eur. J. Clin. Microbiol. Infect. Dis. 23:529–534.

[3] Bickley, J., Hopkins, D. (1999) Inhibitors and enhancers of PCR. In: Analytical Molecular Biology; Quality and Validation. Edited by Saunders, C., Parkes, H.C., Primrose, B. Laboratory of the Government Chemist, Royal Society of Chemistry.

[4] Kosek, M., Bern, C., Guerrant, R.L. (2003) The global burden of diarrhoeal disease, as estimated from studies published between 1992 and 2000. Bull. W.H.O. 81:197–204.

[5] McLauchlin, J., Grant, K.A. (2007): Clostridium. In: Foodborne Diseases. Edited by Simjee, S., Humana Press, Totowa.

12 Pathogens relevant in the central nervous system
Helene Peigue-Lafeuille and Cécile Henquell

The main clinical presentation in patients with infection of the central nervous system (CNS) is meningitis and/or encephalitis. Infection of the CNS can be produced by numerous viruses and bacteria. Many infections of the CNS are potentially life threatening and prognosis largely depends on how quickly specific treatment, if it exists, is started. Hence, the physician's first decision on admission should be to consider the etiology for which a specific treatment must be initiated.

This chapter will focus on CNS acquired infections in immunocompetent adults and children (except neonates). (For HIV infection, see Chapter 7; for infections in immunosuppressed patients including infections produced by herpesviruses, see also Chapter 9.) The viruses involved here are enteroviruses (EV), Epstein-Barr virus (EBV), herpes simplex virus type 1 (HSV-1) and type 2 (HSV-2), human herpesvirus 6 (HHV-6), Nipah virus, parechovirus, rabies virus, tick-borne encephalitis virus (TBEV), Toscana virus (TOSV), varicella zoster virus (VZV), and West Nile virus (WNV). The bacteria involved here are *Anaplasma phagocytophilum, Borrelia burgdorferi, Ehrlichia chaffeensis, Haemophilus influenzae type b, Listeria monocytogenes, Mycobacterium tuberculosis, Neisseria meningitides,* and *Streptococcus pneumoniae*.

The enterovirus (EV) is a single-stranded RNA positive virus and a genus of the *Picornaviridae* family. It is by far the most common CNS pathogen. The etiologic agent of 80–92% of cases with aseptic meningitis is EV. The genus EV is classified into four species, EV-A to EV-D including 105 human-pathogenic viruses. The three poliovirus serotypes belong to the species EV-C. Considerable efforts have been made to eradicate poliomyelitis, which was found in only a small handful of nations, mostly in Africa, India, and Pakistan until 2009–2010 but is re-emerging now, for example in China in 2011 and in Syria in 2013. Clinical syndromes caused by nonpolio EV involving CNS include meningitis, encephalitis, and acute flaccid paralysis. Acute flaccid paralysis and encephalitis are rare events. However, a few serotypes are under surveillance, among them enterovirus 71, which has caused large outbreaks in the Asian Pacific region since 1997, is present now in Europe and is responsible for severe encephalitis with high morbidity and mortality. Benign aseptic meningitis involving coxsackieviruses and echoviruses is by far the most frequent clinical presentation of CNS infections. It occurs with both epidemic and endemic patterns including one or several types, irrespective of age or season. Almost 30% of cases are observed in winter, and, if they are systematically investigated, almost 30% of cases are found in adults (Fig. 12.1). The treatment is only supportive.

The EBV is a ubiquitous gamma-herpesvirus and infects up to 95% of individuals. In immunocompetent patients, EBV has been implicated in meningoencephalitis,

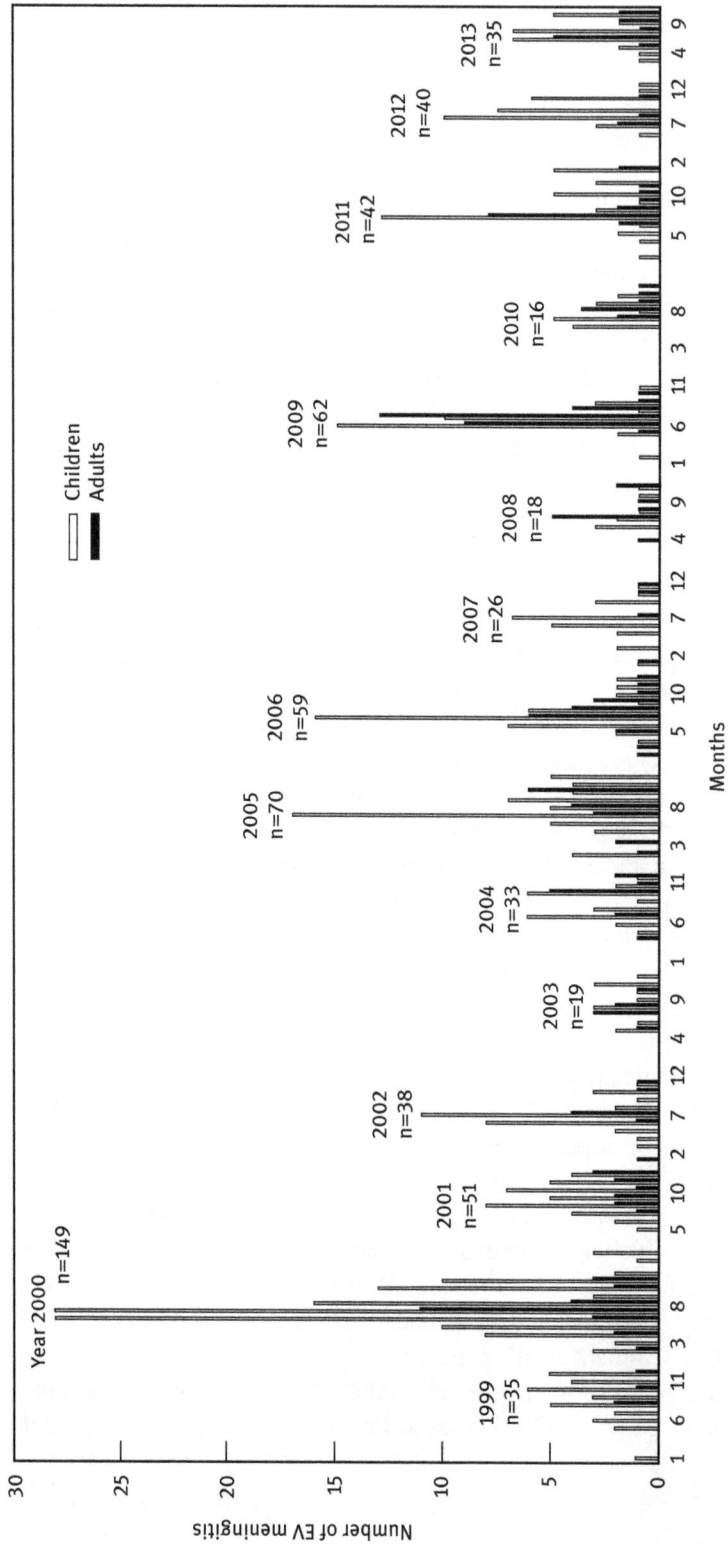

Fig. 12.1: Monthly distribution of enterovirus meningitis prospectively diagnosed by reverse transcription PCR at the teaching hospital of Clermont-Ferrand (France) for children and adults between 1999/01/01 and 2013/10/31.

cerebellar ataxia, and post-infectious encephalomyelitis, either with complicated mononucleosis or as the sole manifestation.

HSV-1 and HSV-2 are ubiquitous alpha-herpesviruses, which infect up to 100% of individuals. Meningitis and meningoencephalitis are rare manifestations, in primary infection as well as in reactivation. However, HSV is the major agent involved in encephalitis with HSV-1 occurring in 95% of cases and HSV-2 in the remaining 5%. The mortality rate in untreated patients is 70%. Consequently, HSV should be considered in all cases of encephalitis and acyclovir should be initiated in all patients prior to results of diagnostic studies.

Infection with HHV6 is very common, with two-thirds of children being seropositive by 1 year of age, as well as 95% of adults. Viral invasion of cerebrospinal fluid (CSF) is a common event during primary infection and detection of HHV6 is positive in >75% of normal brain specimens. Severe encephalitis is a rare complication of HHV6 infection in immunocompetent children.

The Nipah virus is a member of the *Paramyxoviridae* and a single-stranded RNA virus. The genus *Henipavirus* comprises Hendra virus and Nipah virus. Nipah virus first emerged in 1998/1999 in Malaysia, causing encephalitis in adults involved in pig farming or pork production, with a mortality of approximately 40%. Nipah virus has caused several outbreaks in Bangladesh and India with different modes of transmission, mainly person-to-person transmission, potentially food-borne transmission, and rarely nosocomial transmission. The mortality ranges between 40% and 92%, owing to either genetic differences in the strains involved or a lower level of supportive care. The total number of encephalitis cases is low. However, this CNS pathogen is under surveillance.

The parechovirus is a genus of the *Picornaviridae* family. The genus parechovirus includes human parechovirus type 1 and type 2, previously known as echovirus type 22 and type 23, human parechovirus type 3 through type 14, and Ljungan virus, isolated from rodent species. The spectrum of clinical presentations is similar to that of the human enteroviruses and may include aseptic meningitis, encephalitis, acute flaccid paralysis, and neonatal sepsis. Usually, infections are asymptomatic. However, human parechoviruses, especially type 3, may cause severe CNS infection with white matter injury, especially in very young children.

Rabies virus is a single-stranded RNA virus and a member of the *Rhabdoviridae* family. It inevitably causes fatal encephalitis in the absence of postexposure prophylaxis. If rabies is suspected or/and if the patient has died without etiology, special national health reference centers should be contacted.

The TBEV is a single-stranded RNA virus and a member of the genus *Flavivirus* within the family *Flaviviridae*. TBEV may cause fatal encephalitis. There has been a steady increase in cases of tick-borne encephalitis in Europe in the last 30 years, possibly caused by an expanding tick population promoted by factors including climate change. In contrast, a number of European countries have observed a decrease in the number of cases owing to successful vaccination campaigns.

The TOSV is an arbovirus belonging to the *Bunyaviridae* family that is transmitted by the bite of sandflies (*Phlebotomus*). The viral genome consists of three single-stranded RNA segments referred to as S (small), M (medium), and L (large) segments. TOSV recently emerged in Italy and thereafter in Spain, Portugal, and other European countries as a cause of CNS disease, mainly meningitis in summer. It is mainly observed in travelers returning from endemic areas.

The VZV is an alpha-herpesvirus and can cause both meningitis and encephalitis. In 50% of cases, there is no detectable rash. In large surveys, VZV is the second most frequent cause (after HSV) of encephalitis. As in HSV infection, acyclovir must be started immediately.

The WNV is a single-stranded RNA virus and a member of the genus *Flavivirus* within the family *Flaviviridae*. It is the most dramatic example of a virus spreading into a new host range. WNV emerged in the USA in 1999. It was transmitted from birds to humans primarily by the *Culex* species of mosquitoes. It is responsible for CNS disease with a fatality rate between 7% and 10%. Meanwhile, other modes of WNV transmission were recognized including transfusion of blood and blood products, solid organ transplantation, and occupational exposure in laboratory workers.

Anaplasma phagocytophilum (formerly *Ehrlichia phagocytophilum*) is transmitted through the bite of infected *Ixodes* ticks and is responsible for human granulocytic anaplasmosis (HGA). The infection occurs in parts of the USA and Europe where Lyme disease is endemic. In contrast to Lyme disease, HGA is infrequently diagnosed in children. Severe CNS manifestations have been reported. All symptomatic patients suspected to have HGA should be treated with antimicrobial therapy (doxycycline) because of the risk of complications.

Borrelia burgdorferi is a spirochete responsible for Lyme disease. Lyme disease seems to be an emerging disease; it is estimated that approximately 30% of ticks are contaminated in France and up to 60% in Austria.

Ehrlichia chaffeensis is transmitted by tick bites and is responsible for human monocytotrophic ehrlichiosis with potentially severe CNS manifestations. As in HGA, treatment is doxycycline.

Haemophilus influenzae type b (Hib) is responsible for acute community-acquired bacterial meningitis. The introduction of vaccination against bacterial meningitis agents has modified disease epidemiology, especially in developed countries. The introduction of vaccination against Hib was followed by a dramatic decrease in meningitis. It is assumed that it also decreased infection by reducing nasopharyngeal carriage of the bacteria.

Listeria monocytogenes is responsible for acute community-acquired bacterial meningitis. Except in neonates, *Listeria monocytogenes* meningitis occurs almost exclusively in patients older than 50 years of age.

Mycobacterium tuberculosis CNS disease is uncommon but severe. Several risk factors for CNS tuberculosis have been identified. Children, HIV-coinfected patients, and people who live in unfavorable socioeconomic conditions are at higher risk of

CNS tuberculosis. Other risk factors include malnutrition, recent measles, alcoholism, malignancies, and immunosuppressive therapy. The re-emergence of measles in many developed countries in recent years is another factor to be taken into consideration. In developed countries, foreign-born individuals are overrepresented among CNS tuberculosis cases.

Neisseria meningitidis and *Streptococcus pneumoniae* are the two most common pathogens, in infants (>4 weeks) and children, and in adults. These two agents represent more than 80% of all cases and 95% in patients from 1 to 24 years old. *Streptococcus pneumoniae* accounts for 70% of cases in adults older than 40 years of age, followed by *Neisseria meningitidis* and more rarely *Listeria monocytogenes* and *Haemophilus influenzae* (5–10% of cases for each). The introduction of a pneumococcal conjugate vaccine in the United States was followed by a decrease in the rate of infection by nearly 60% in children <5 years of age. Mortality at the acute stage remains high, approximately 20% in adults, and 10% in children, especially for pneumococcal meningitis and invasive meningococcal meningitis with bacteremia. Approximately 30–50% of survivors suffer from permanent sequelae.

12.1 Clinical manifestations

Table 12.1 shows the areas of infection and clinical manifestations associated with major CNS pathogens.

12.1.1 Viral meningitis

Benign aseptic meningitis is by far the most frequent clinical presentation of CNS infections and EVs are involved in approximately 90% of cases. Although the initial presentation may mimic CNS infections produced by other pathogens (especially bacteria and herpesviruses), a major difference is that EV meningitis does not need any treatment or additional investigations. The disease is benign, with usually a short course and no sequelae. When patients are managed by experienced physicians familiar with EV meningitis, the length of stay in hospital does not exceed 2 days. In typical cases, children are often not hospitalized at all or immediately discharged after EV RNA has been rapidly detected, resulting in significant healthcare savings.

12.1.2 Acute community-acquired bacterial meningitis

In acute bacterial meningitis (ABM), the meningitis syndrome is the clinical "gold standard" for clinical diagnosis. However, the classic triad of fever, neck stiffness, and a change in mental status is not present in almost 50% of patients. Ninety-five

Tab. 12.1: Epidemiology and clinical manifestations associated with major CNS pathogens in immunocompetent patients except neonates.

	Epidemiology	Clinical manifestation	Treatment/vaccine
VIRUS			
EV	Worldwide distribution; peak incidence in late summer and early fall, 30% of cases in winter	Aseptic meningitis extremely frequent, rarely encephalitis and acute flaccid paralysis, enterovirus 71: rhombencephalitis and polio-like syndrome	Supportive
EBV	Exposure to saliva of infected person	Seizures, coma, cerebellar ataxia, cranial nerve palsy	Supportive
HSV	All age groups, all seasons	Encephalitis, behavioral abnormalities, memory impairment	Acyclovir urgently in all cases of suspicion
HHV6	Rare in the immunocompetent, no seasonal predilection	Exanthema subitum, seizures	Gancyclovir, Foscarnet
Nipah virus	Close exposure to infected pigs, bats, infected humans, South Asia	Fever, headache, altered mental status, myoclonus, dystonia, areflexia, pneumonitis	Supportive
Parechovirus	Worldwide, very young children under 3 months of age	Meningitis, encephalitis, neonatal sepsis	Supportive
Rabies virus	Bite of infected animal, few recognized cases by organ donation, rare in developed countries, travelers at risk, worldwide distributed	Encephalitis, furious form leading to stupor, coma and death, paralytic form with ascending paralysis	Only supportive after onset of disease, post-exposure prophylaxis (immunoglobulin and vaccine), vaccine
TBEV	Tick vector, rodent reservoir, unpasteurized milk, Eastern Russia, Central Europe, Far East, epidemic, sporadic	Encephalitis, poliomyelitis-like paralysis	Supportive, vaccine
TOSV	Bite of sandflies (*Phlebotomus*), Italy, Spain, Portugal, Europe, travelers	Self-limited febrile illness, meningitis with tremor, myalgia, rarely severe encephalitis	Supportive
VZV	All age groups, highest in adults, all seasons, rash may be absent	Chickenpox, zoster, encephalitis with focal neurologic deficit and seizures, pneumonia	Acyclovir, Gancyclovir, urgent in cases of suspicion, pending the results, vaccine

(Continued)

Tab. 12.1: (*Continued*)

	Epidemiology	Clinical manifestation	Treatment/vaccine
WNV	Mosquito vector, bird reservoir, North and Central America, Africa, Asia, middle East and southern Europe, Italy, Macedonia, transmission reported via transfusion, transplantation	Abrupt onset of fever, headache, neck stiffness and vomiting, meningitis, encephalitis, acute flaccid paralysis	Supportive
BACTERIA			
Anaplasma phagocytophilum[a]	Tick vector, Europe, mid-Atlantic, northern USA	Rare CNS infection, abrupt onset of fever, headache, myalgia, altered mental status	Specific antibiotics, not to be delayed if suspicion
Borrelia burgdorferi[a]	Tick vector, North America, Europe, Asia	Early and shortly after erythema migrans or late with arthritis, facial nerve palsy, meningitis, radiculitis, papilledema	Specific antibiotics
Ehrlichia chaffeensis	Tick vector, parts of USA	Rare CNS infection, confusion, photophobia, stupor, hallucinations, seizures, coma	Specific antibiotics
Haemophilus influenzae	Worldwide, decreasing numbers because of vaccination	Acute meningitis	Specific antibiotics, vaccine
Listeria monocytogenes	Adults >50 years of age, alcohol abuse	Rhombencephalitis with ataxia, cranial nerve deficits, nystagmus	Specific antibiotics
Mycobacterium tuberculosis	More common in developing countries, risk factors include unfavorable socioeconomic conditions, malnutrition, recent measles, alcoholism, malignancies	Chronic meningitis with insidious onset, encephalitis, many atypical cases in recent studies involving young adults with semi-rapid onset	Specific antibiotics, vaccine
Neisseria meningitidis	Worldwide with different distribution according to serogroups, sporadic outbreaks, large epidemics	Acute meningitis, rash, invasive forms with bacteremia, high mortality	Specific antibiotics prior (hospitalization in case of invasive form), vaccine for certain serogroups
Streptococcus pneumoniae	Worldwide with different distribution according to serotypes	First cause of acute meningitis, upper respiratory tract infection (otitis, sinusitis), pneumonia	Specific antibiotics, vaccine for certain serotypes

[a] The possibility of coinfection with *Borrelia burgdorferi* or *Anaplasma phagocytophilum* in patients with severe or persistent symptoms must be considered.

Prior to admission:
Antibiotics if invasive Neisseria meningitidis disease

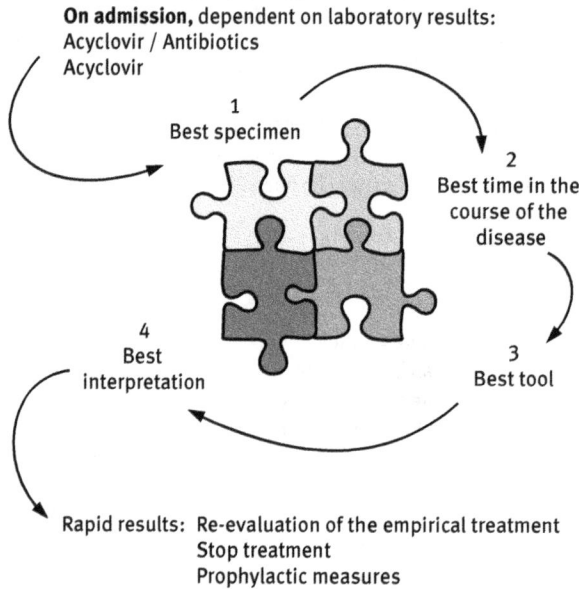

On admission, dependent on laboratory results:
Acyclovir / Antibiotics
Acyclovir

1
Best specimen

2
Best time in the
course of the
disease

4
Best
interpretation

3
Best tool

Rapid results: Re-evaluation of the empirical treatment
Stop treatment
Prophylactic measures

Fig. 12.2: General suggestions for clinical use of molecular tests in CNS infectious disease.

percent of patients have at least two of the four symptoms of headache, fever, neck stiffness, and altered mental status. The level of consciousness ranges from somnolence and confusion to coma. At the late stage, the terms "encephalitis" or "meningoencephalitis" are more appropriate than "meningitis". In very young children, family physicians should be alert to any signs of sepsis. Antibiotics before hospitalization should only be initiated for patients with suspicion of disseminated *Neisseria meningitidis* infection (meningococcemia) with purpuric rash. All patients with suspected ABM should be hospitalized as soon as possible for rapid assessment and treatment. Lumbar puncture for examination of CSF using the most appropriate diagnostic tools should be widely performed because of the significant mortality and morbidity of the disease (Fig. 12.2).

12.1.3 *Mycobacterium tuberculosis*

CNS disease caused by *Mycobacterium tuberculosis* is an uncommon but highly devastating manifestation of tuberculosis. It shows high mortality and morbidity. Early treatment can dramatically improve outcomes. It is important to consider *Mycobacterium tuberculosis* as a potential pathogen of CNS disease. CNS involvement has been reported in 6.3% of extrapulmonary tuberculosis cases representing 1.3% of total tuberculosis cases.

12.1.4 Encephalitis

Encephalitis may be caused by more than a hundred pathogens with viruses predominating. Among them, herpesviruses are the major agents. However, if the incidence of these infections in the general population is considered, encephalitis remains a rare complication.

In the California Encephalitis Project 1998–2005, a causative pathogen was found in approximately 30% of cases including, in decreasing order, EV, HSV-1, VZV, WNV, *Mycobacterium tuberculosis*, EBV, and pyogenic bacteria. In the 2007 French study, a causative organism was shown in 52% of cases. The main agents identified were, in decreasing order, HSV-1, VZV, *Mycobacterium tuberculosis*, and *Listeria monocytogenes*, and they represented >80% of cases in which an organism was identified.

Similar to acute bacterial meningitis and as a rule, the first step is to consider an organism for which there is a specific treatment, the absence of which will be harmful for the patient, and rapidly use the best diagnostic tools, especially molecular tools to approach or reach this goal (Fig. 12.2). This procedure is of crucial importance for HSV and VZV on the one hand and acyclovir on the other hand.

12.2 Preanalytics

12.2.1 Goals of etiological investigations

For the patient, determining the etiology of the CNS infection is critical in providing therapeutic guidance, initiating specific treatment, and reconsidering or stopping empirical treatment. Full clinical information should be provided to the laboratory, including particular suspicions, antibiotics administered before lumbar puncture (if there is a strong suspicion of meningococcal infection or a delay exceeding 90 min in hospital transfer), and antiviral therapy (acyclovir in the case of HSV and VZV suspicion). This is crucial for the interpretation of specific tests (Fig. 12.2).

It is important to take prophylactic measures for the patient's immediate family and acquaintances (especially in cases of *Neisseria meningitidis* and *Mycobacterium tuberculosis* suspicion) and to take or reinforce hygiene measures to avoid outbreaks.

If a given pathogen is frequently involved, the corresponding disease is benign, and the treatment only supportive (in the case of EV meningitis), a rapid prospective diagnosis could save health costs by stopping unnecessary investigations or neuroimaging studies and reducing the length of hospitalization and antibiotic therapy.

12.2.2 Specimens and handling

Specimens useful for detection of nucleic acids may be collected from different body sites with CSF being the ideal specimen in almost all cases (Tab. 12.2).

Tab. 12.2: Pathogen related specimens.

	Specimen
VIRUS	
EV	CSF (sometimes together with throat swab and stool)
EBV	CSF
HSV	CSF
HHV6	CSF, EDTA whole blood
Nipah virus	CSF
Parechovirus	CSF
Rabies virus	CSF, saliva, nuchal biopsy
TBEV	CSF
TOSV	CSF
VZV	CSF
WNV	CSF
BACTERIA	
Anaplasma phagocytophilum	CSF, whole blood specimen
Borrelia burgdorferi	CSF, skin biopsy
Ehrlichia chaffeensis	CSF, whole blood specimen
Haemophilus influenzae	CSF
Listeria monocytogenes	CSF
Mycobacterium tuberculosis	CSF
Neisseria meningitidis	CSF, skin biopsy
Streptococcus pneumoniae	CSF

12.2.3 Time of lumbar puncture during the course of illness and quantity of CSF required

Lumbar puncture is generally performed at patient admission. If not, lumbar puncture should be performed within 1–2 days following the onset of neurological symptoms, a time at which the amount of replicating virus is thought to be at its peak. The time at which lumbar puncture is performed in the course of the disease is an important element that the laboratory needs for postanalytic interpretation (Fig. 12.3).

The quantity of CSF should not be less than 2 ml (30–40 drops) in children and at least 5 ml (100 drops) in adults. If *Mycobacterium tuberculosis* is suspected, more fluid is required. The first CSF sample on admission is the best indicator of brain infection and therefore the best specimen to demonstrate the presence of a given pathogen. Numerous agents may be involved. As they are not always suspected in the first batch of investigations, exhaustive retrospective examinations often have to be made of a stored specimen. The following factors should be taken into consideration: (1) sometimes very complex clinical situations in practice, (2) potential severity of the disease, (3) number of undetermined causes in CNS infections (especially encephalitis), and (4) need for further investigations for some emerging diseases with CNS symptoms. Consequently, every effort has to be made not to waste even small amounts of the initial CSF on unnecessary or nonrelevant diagnostic tests.

Case A

Case B

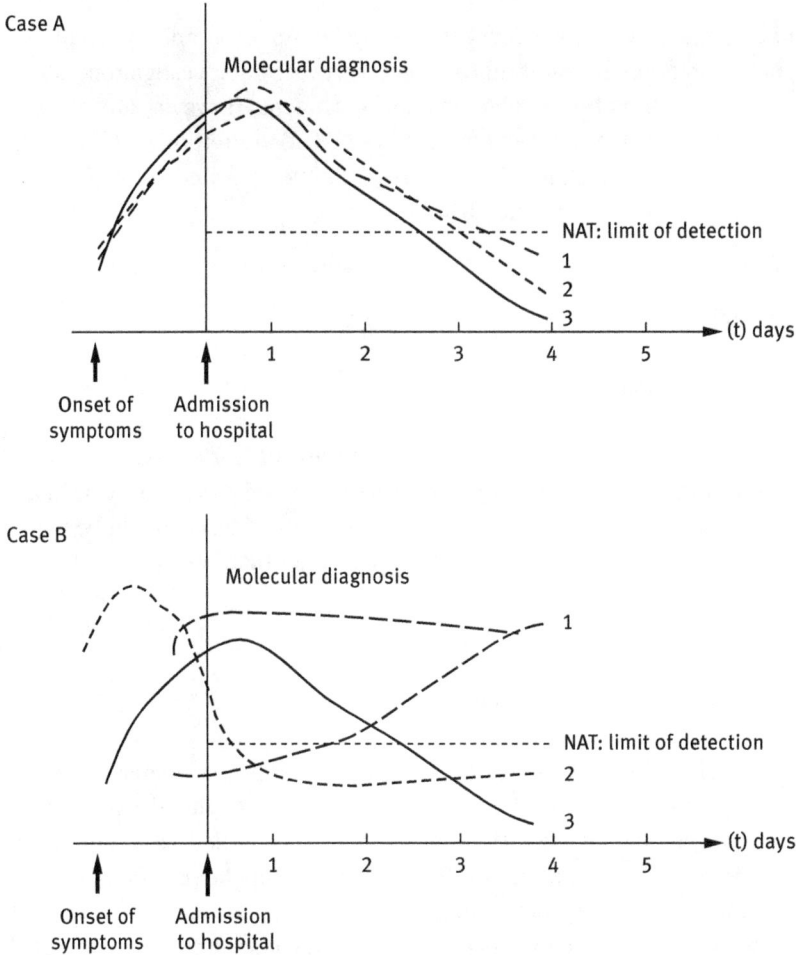

Fig. 12.3: Reliability of molecular diagnostics and kinetics of pathogens in cerebrospinal fluid (CSF). In case A, nucleic acid testing (NAT) is useful for the 3 pathogens considered. In case B, NAT is useful only for pathogen 3 and in certain cases for pathogen 1. For pathogens 1 and 2, DNA or RNA may not be detectable in CSF obtained on admission. Pathogen 1 could be herpes simplex virus as HSV DNA may be detectable either in CSF on admission or only 3 to 7 days after onset of neurological symptoms. Pathogen 2 could be West Nile virus as the maximum RNA concentration in CSF is found before occurrence of symptoms. Pathogen 3 could be enterovirus infection.

12.2.4 Transport and storage of specimens

Rapid transport at room temperature to the laboratory is mandatory, so that first results (cell count, Gram staining) are communicated within an hour after lumbar puncture. For molecular techniques, CSF is centrifuged 3 min at 1500 rpm to eliminate red blood cells and impurities such as heme and potentially endogenous polymerase inhibitors. The sensitivity of molecular assays is related to the volume of the specimen tested and to bacterial or viral load. Most assays can be run on samples of 30 µl, but 100–200 µl may yield optimal results and are necessary for viral investigations. Molecular assays are best run on freshly obtained CSF specimens. However, if necessary, an aliquot should be stored rapidly (within 24 h) at –20°C for a short time. If long-term storage is anticipated, several aliquots of specimens should be stored at –70°C. Multiple freeze–thaw cycles should be avoided.

12.3 Analytics

12.3.1 Sample preparation

Automated platforms have largely replaced manual protocols (see also Chapter 4). In CNS infections, results of molecular diagnostic assays should be available within the time frame of clinical decision-making. Therefore, on-demand tests should be possible on the arrival of CSF specimens at the laboratory rather than waiting for the next large series.

12.3.2 Nucleic acid amplification and detection

There are commercially interesting pathogens and those of no commercial interest. In many cases, laboratory-developed assays using either conventional PCR or more frequently qPCR are commonly used. Their reliability may be sometimes even greater than that of commercial assays, as shown in results of *Quality Control for Molecular Diagnostics* (QCMD) panels (www.qcmd.org).

For EV, molecular assays are directed to amplify the conserved 5'-nontranslated conserved region in the RNA genome of all enteroviruses. Numerous techniques and tests are available: assays that detect PCR products in microtiter wells, those using qPCR, and recently a fully automated multiplex RT-qPCR assay combining automated nucleic acid sample preparation, amplification, and real-time detection of EV RNA. A summary of currently frequently used commercial assays in Europe is provided in Tab. 12.3. In the 2013 QCMD EV program, 92% of laboratories used RT-qPCR assays; 52% of them were laboratory-developed assays.

Regarding EBV, see Chapter 9.

For HSV, a summary of currently frequently used commercial assays in Europe is provided in Tab. 12.4. In the 2013 QCMD HSV program, 92% of laboratories used qPCR assays; 50% of them were laboratory-developed assays.

Regarding HHV-6, see Chapter 9.

For Nipah virus, there is currently no commercial test available. RT-qPCR has been developed using primers for the Nipah NP gene and a TaqMan probe for detection.

Tab. 12.3: Comparison of currently frequently used commercially available assays for the qualitative detection of EV RNA.

Characteristics	Manufacturer and details				
	AnDiaTec	Argene	Cepheid AB	Nanogen/ ELITechGroup	Qiagen GmbH
Kit name	Enterovirus RT-PCR Kit	Enterovirus R-gene	Xpert EV	Enterovirus Q-PCR Alert Kit	artus Entero- virus RT-PCR Kit
Target sequence	5'-noncoding region	5'-noncoding region	5'-noncoding region	5'-noncoding region	5'-noncoding region
Amplification method	qPCR	qPCR	qPCR	qPCR	qPCR
Detection method	Fluorescence	Fluorescence	Fluorescence	Fluorescence	Fluorescence
Internal control	Heterologous	Heterologous	Heterologous	Heterologous	Heterologous

The technique allows rapid detection and quantitation of Nipah RNA in both field and experimental materials used for surveillance and for specific diagnosis.

For parechovirus detection, molecular assays for EV detection are not useful. An RT-qPCR based laboratory-developed assay was described recently for detection of all known members of the genus *Parechovirus*. A commercial assay is now available. Molecular testing for parechovirus is included in the QCMD program since 2010. The assay targets the conserved region in the 5'-nontranslated region (5'-NTR) of the parechovirus genome and can detect both human parechoviruses and Ljungan virus.

For rabies virus, RT-qPCR was recently developed to improve timely diagnosis of ante-mortem human rabies. Three sets of two primers and one internal dual-labeled probe for each primer set that targets distinct conserved regions of the rabies virus *N* were assessed.

For TBEV, an RT qPCR including an internal control was developed. The assay targets a fragment of the 3'-noncoding region of the TBE genome.

For TOSV, laboratory-developed molecular assays using qPCR with TaqMan probe detection were introduced recently. The *S* segment and a fragment of the *N* gene were used as the target. Assays were able to detect two lineages of a TOSV endemic in Italy and Spain.

Regarding VZV, see Chapter 9.

For WNV, several assays based on RT qPCR have been developed. The non-structural NS2 and NS5 proteins are used as the target. Recently, a multiplex reverse transcriptase-PCR ligase detection assay was developed for the detection of WNV in both clinical and mosquito pool samples amplifying three different genomic regions in nonstructural proteins NS2a and NS5 in order to minimize the risk of detection failure owing to genetic variation.

Tab. 12.4: Comparison of currently frequently used commercially available assays for the qualitative detection/quantitation of HSV DNA.

Characteristics	Manufacturer and details				
	Argene	**Cepheid**	**Nanogen/ELITechGroup**	**Qiagen**	**Roche**
Kit name	HSV1 HSV2 VZV R-gene	affigene HSV 1/2 tracer	HSV 1, HSV 2 ELITe MGB Kit	artus HSV-1/2 PCR Kit	LightCycler HSV 1/2 Qual Kit
Target sequence	US7 (glycoprotein I, HSV-1); US2 (hypothetical protein, HSV-2)	US6 (glycoprotein D, HSV-1); US4 (glycoprotein G, HSV-2)	US 6 (glycoprotein D, HSV-1) and US4 (glycoprotein G, HSV-2)	UL27 (glycoprotein B)	UL30 (DNA polymerase)
Amplification method	qPCR	qPCR	qPCR	qPCR	qPCR
Detection method	Fluorescence	Fluorescence	Fluorescence	Fluorescence	Fluorescence
Internal control	Heterologous	Heterologous	Heterologous	Heterologous	Heterologous
Standards	Four EQS	–a	Four EQS	Four EQS	–a
Analytic measuring range	2.8×10^2–1.5×10^5 copies/ml (HSV-1); 7.0×10^1–1.0×10^6 copies/ml (HSV-2)	–b	1.25×10^2–1.25×10^6 copies/ml	NA	–c

a Qualitative test.
b Limit of detection 1.3×10^2–1.8×10^2 copies/ml (depending on the sample preparation protocol employed).
c Limit of detection 4×10^2 copies/ml.
EQS: external quantitation standards; NA: data not available.

To evidence either viruses of different families or different virus types in the same family, multiplex PCR assays are in development. The simultaneous detection of herpesviruses (DNA polymerase gene), enteroviruses (5′-noncoding region), and five flavi-viruses (NS5 region) using amplification by PCR and detection of amplification products using DNA microarray technology was recently reported. Furthermore, multiplex RT-qPCR has been developed for the detection of eight medically important flaviviruses in mosquitoes and for clinical diagnosis. The latest development, which will perhaps allow considerable improvement in the diagnosis of the *Flavivirus* genus, is the use of universal primers that amplify an 800-bp section of the NS5 gene from all (three) major flavivirus subgroups followed by sequencing. Such an assay allows detection of several flaviviruses including WNV, Dengue viruses 1, 2, and 4, Japanese encephalitis virus, Murray valley encephalitis virus, St Louis encephalitis virus, TBEV, and Yellow Fever virus. While the analytical sensitivity of these new tests is good, their clinical sensitivity has yet to be assessed. Figure 12.3 shows examples of multiplex results and reliability.

For *Anaplasma phagocytophilum*, nested PCR has been used to amplify the 16S rRNA gene from *Anaplasma phagocytophilum*.

For *Borrelia burgdorferi*, nested PCR has been used to amplify the flagellin gene from *B. burgdorferi*. qPCR assays have been developed, using *RecA* gene as the target.

For the simultaneous detection of the *Anaplasma phagocytophilum msp2* gene and *Borrelia burgdorferi* 23S rRNA, a multiplex qPCR assay was developed. The assay was tested on various *Anaplasma*, *Borrelia*, *Ehrlichia*, and *Rickettsia* species and was found to be highly specific for *Anaplasma phagocytophilum* and the *Borrelia* species tested.

For *Ehrlichia chaffeensis*, qPCR has been developed to amplify the 16S rRNA gene from *Ehrlichia chaffeensis*.

For *Haemophilus influenzae* detection, a qPCR multiplex assay has been developed targeting the capsulation *bex*A gene, the *ompP2* genes, and 16S rRNA.

For *Listeria monocytogenes*, molecular assays using qPCR have been used, mainly in studies of food production. However, multiplex assays including *L. monocytogenes* detection have been reported with 16S rRNA as a target in clinical situations.

For *Mycobacterium tuberculosis*, there are several different techniques and targets in use, both commercial and noncommercial. The challenge for molecular diagnosis is the low number of bacteria and the presence of amplification inhibitors in CSF. The commercial molecular tests for respiratory specimens have not been approved for CSF. PCR assays have been performed with the use of primers directed at the insertion sequence *IS6110*, a repetitive element exclusively found in the genome of the *Mycobacterium tuberculosis* complex. However, inconsistent results have been obtained. Commercially available and other molecular tests may not provide a perfect screening tool for the diagnosis of *Mycobacterium tuberculosis*.

For *Neisseria meningitidis*, the 16S rRNA has been used as the target in several assays. qPCR using the capsular transport *ctrA* gene as the target was developed recently for diagnosis and for serogrouping in a prospective study of meningococcal disease in children.

For *Streptococcus pneumoniae*, qPCR based laboratory-developed assays have been used targeting either the pneumolysin *ply* gene or the *lytA* gene.

To detect the causative pathogen in CSF specimens collected from patients with bacterial meningitis, multiplex PCR assays have been developed, the majority of them based on qPCR. One of these assays is able to detect the three most frequent pathogens, *Neisseria meningitidis* (via the capsular transport *ctrA* gene), *Haemophilus influenzae* (via the capsulation *bexA* gene), and *Streptococcus pneumoniae* (via the pneumolysin *ply* gene). Another was designed to detect eight bacterial CSF pathogens. Clinical evaluations of these assays are not yet available.

For detection of the 16SrRNA gene of Eubacteria in general, universal PCR can be used. A positive result indicates the presence of bacteria in specimens from usually sterile locations such as CSF, and the bacterial species can be identified by subsequent sequencing. Sensitivities range from 59% to 93% according to studies, specificities range from 97% to 98%. For use in the routine diagnostic laboratory, the sensitivity of universal PCR has to increase and standardization is needed.

12.4 Postanalytics

Table 12.5 shows the interpretation of results obtained with molecular assays for detection of CNS pathogens, and the status of molecular tests in microbiological diagnostics.

12.4.1 Workflow and testing schedules for molecular tests

For the optimal management of patients, molecular diagnosis results must be quickly available (Fig. 12.3). Workflow and testing schedules for molecular tests are affected by the time taken for specimen transportation, the arrival times of specimens at the laboratory, the clinical urgency of the results, and the laboratory's hours of operation. For example, molecular tests for HSV and EV are only useful for the optimal management of patients if they are available together with hospitalization. The most important factors in postanalytics are when and how a positive result of a molecular test is taken into consideration by the physician in charge of the patient. Once a positive result is obtained, the patient must be re-assessed and a decision taken. If the molecular test is not performed in the local laboratory, a rapid result is not gained. Likewise, if it is performed in the local laboratory during after-hours but not taken into consideration until the next working day, no time will be saved. It may be most efficient to implement the test daily (7 days a week) and/or with a more rapid turn-around time.

Tab. 12.5: Interpretation of results obtained with molecular assay for CNS pathogens.

	Molecular assay
VIRUS	
EV	Gold standard, sensitivity 96–100%, specificity 96% compared with cell cultures, rapid result, if positive reconsider clinical data and reduce medical cost; sensitivity may be lower in EV71 and poliovirus infections because of low viral load in CSF; additional throat swabs and/or stool specimens may be helpful
EBV	Quantitative results are helpful
HSV	Gold standard, sensitivity >95%, specificity >99%, rapid results, Acyclovir dependent on the results, if negative and clinical symptoms or positive neuro-imaging, repeat the test 3–7 days later and maintain Acyclovir
HHV6	High rate of detection in healthy adults (latent virus) and poor positive predictive value (comparison with EDTA whole blood viral load required)
Nipah virus	Refer to special laboratory, do not replace serologic testing
Parechovirus	Specific PCR useful in EV-like disease, emergent or underestimated etiology
Rabies virus	PCR on skin biopsy, CSF, and saliva, refer to special laboratory
TBEV	Useful at the onset (before day 4), after day 4, false-negative (serology!)
TOSV	Useful at the onset before day 5, after day 5, false-negative (serology!)
VZV	Gold standard (with IgM in CSF)
WNV	Useful at the onset, before day 4, after day 4 false-negative (serology – with IgM in CSF)
BACTERIA	
Anaplasma phagocytophilum	PCR of whole blood specimen prior to initiation of antibiotic therapy (not to be delayed), with smear examination, maybe false-negative PCR of CSF: low yield (serology!)
Borrelia burgdorferi	PCR of skin biopsy (erythema migrans) with maximum sensitivity and specificity, PCR of CSF with low sensitivity (17% in neuroborreliosis), useful in patients with a short duration (<14 days) of disease (serology!)
Ehrlichia chaffeensis	PCR with low sensitivity (serology!)
Haemophilus influenzae	Culture gold standard, additional cases detected with PCR of CSF in culture-negative cases
Listeria monocytogenes	Culture gold standard, additional cases detected with PCR of CSF in culture-negative cases
Mycobacterium tuberculosis	PCR often with poor sensitivity (33–91% in various studies), consider complete clinical and epidemiological data for diagnosis
Neisseria meningitidis	Culture gold standard, additional cases detected with PCR of CSF in culture-negative cases, PCR in skin biopsy useful
Streptococcus pneumoniae	Culture gold standard, additional cases detected with PCR of CSF in culture-negative cases

Tab. 12.6: Interpretation of results and limitations of molecular tests in CNS disease.

False-negative results	False-positive results
Transportation problem	Contamination
Inhibition	Nonspecific amplification
Mutations	Sequence homology among different bacteria
Primer binding site genetic change	Database problem
Database problem	Amplification of latent virus in PBMCs
Low concentration of infectious pathogen	Breakdown in the blood-brain barrier
Final assay volume	Contamination of the CSF with blood
Inappropriate moment in the course of the illness	
Local immune response	

12.4.2 Limitations of molecular tests

Molecular tests have limitations that should carefully be taken into consideration (Tab. 12.6).

12.4.2.1 Negative molecular detection of a given pathogen in CSF does not necessarily rule out a specific agent

Problems may occur during transportation and cause nucleic acid degradation. Inhibitors such as endo- and exonucleases are much less commonly present in CSF than in other body fluids or tissues. However, false-negatives are possible even if inhibitor controls are adequately incorporated in the assay: in the acute phase of infection, little is known about the viral load of many agents; in the course of an established infection, little is known about the time taken for clearance and the possible interference of general and local immune response (Fig. 12.3). For example, HSV DNA may be undetectable at a very early stage of the disease. Furthermore, in VZV CNS disease, even despite the presence of IgM antibodies in CSF, VZV DNA may be undetectable. A negative molecular result thus cannot be used as definitive evidence against the diagnosis.

12.4.2.2 Positive molecular detection of a given pathogen in CSF does not necessarily have clinical significance

The clinical significance of a positive result can be clarified through clinical and epidemiological background information combined with quantitative information on the pathogen present. For example, EBV has been implicated in CNS diseases in immunocompetent patients. Seropositivity is approximately 90% in adults. If EBV DNA is detected in CSF, a quantitative study should be performed to differentiate active infections from latent infections. Likewise, detection of HHV6 DNA in CSF should be interpreted with caution in immunocompetent patients. It may correspond to a chromosomal integration or the amplification of a latent virus and the detection of HHV6 DNA is probably not related to the patient's encephalitis. Thus, detection of HHV6 DNA has a poor positive predictive value of less than 30%. Viral load in EDTA whole blood and in CSF must be compared to distinguish between latent and replicating viruses.

12.4.3 Viral CNS infections

Classic biochemical markers in CSF (cell count, lymphocyte/polymorphonuclear cell ratio, proteins) may be lacking. Cultures should be abandoned because of poor sensitivity and the length to result. Molecular tests should thus be performed systematically for the investigation and management of viral CNS infections. Exclusion of viral infections has the advantage that unnecessary therapeutic options – especially antibiotics – are not implemented. Rapid and specific viral diagnosis is thus of relevance in the fight against antibiotic resistance.

Molecular detection of EV RNA to supersede further investigations and HSV DNA because of the severity of CNS infection is of major importance. As recommended in standard guidelines, acyclovir should be initiated in all patients with suspected encephalitis prior to results of diagnostic studies. Detection of EV RNA in the CSF specimen allows a definite diagnosis avoiding further laboratory and neuro-imaging investigations and unnecessary hospitalization and treatment. If HSV DNA is undetectable in CSF, but the suspicion of HSV infection persists, the molecular assay should be repeated on another CSF specimen 3–7 days later. Acyclovir should be continued because the probability of undetectable HSV DNA in a CSF specimen obtained early (less than 72 h) after onset of symptoms in patients in whom the clinical suspicion of HSV encephalitis remains high.

For TBEV and TOSV infections, serological testing remains of interest. Molecular detection on CSF specimens and, more interestingly, on skin biopsies at the site of rash can provide positive results only at an early stage of the disease, that is, in the first 4 or 5 days, and add to the serological results.

12.4.4 Bacterial CNS infections

Gram stain and culture are the standard tests. Gram stain results should be obtained within 1 h and communicated to the physician in charge of the patient. Cultures are always required for bacterial CNS infections and remain the gold standard. However, molecular tests allow detection of additional positives in culture-negative samples for *Neisseria meningitidis* and *Streptococcus pneumoniae*, which are the two most frequent bacteria, and *Haemophilus influenzae* in the following situations: (1) the patient was on antibiotics before lumbar puncture, (2) a critical clinical situation exists, (3) lumbar puncture was delayed due to logistics or transport problems. Molecular testing helps to reduce the overall cost of patient care and creates a rational use of antibiotics, which will subsequently help to decrease resistance problems.

Mycobacterium tuberculosis infection requires particular attention. Early diagnosis and specific treatment correlate with better outcome. However, early diagnosis is difficult because of several issues including numerous atypical features confounding early diagnosis, the insensitivity of traditional staining, the length of *Mycobacterium tuberculosis* culture that should be systematically performed if infection is suspected, the gradual disappearance of acid-fast bacillus staining in this "old" disease, and the need to improve the sensitivity and performance of molecular diagnostic tools. Thus, the management of CNS tuberculosis remains a formidable challenge. With any diagnostic test for tuberculosis, a negative result cannot exclude the possibility of tuberculosis and clinical judgment remains of paramount importance, particularly since CNS tuberculosis seems to be rather common. Molecular tests for *Mycobacterium tuberculosis* need to be greatly improved. However, if molecular diagnosis fails, molecular typing of Mycobacteria isolated from culture may provide identification within hours rather than weeks and rifampicin and/or isoniazid resistance can be diagnosed up to 2 weeks earlier than current methods.

12.4.5 Which pathogens should we look for?

Despite the use of new techniques including molecular, 50–70% of cases of encephalitis remain of unknown etiology. What should be considered to achieve a most meaningful result?

The spectrum of pathogens may vary from one area to another and from one country to another. WNV infection is one example in the USA. In France, TBEV circulation is limited to German and Swiss border regions, and that of TOSV to southern Europe, where the mosquito *Culex albopictus* is present. Rapid spread of disease agents is increasing between continents. Since 2010–2011, cases of WNV encephalitis have been described in the East of Europe, the South of Italy, and Macedonia. RT-PCR using primers capable of amplifying and subsequent sequencing for identification of almost each recognized member of the genus *Flavivirus* will be a significant improvement in this context.

Enterovirus 71 is regularly observed in Europe with sporadic encephalitis in France, Portugal etc. Reference laboratories should therefore continue culture of the virus and identify isolates using molecular techniques including sequencing and phylogenetic analysis as part of public health policy.

The risk of the re-emergence of poliomyelitis in Europe, after an 11-year period of eradication, is a serious threat, owing to the circulation of the wild-type poliovirus in the Middle East and Africa because of armed conflicts, population displacement, and poverty.

12.5 Conclusion

Molecular diagnosis is of major importance for the investigation of CNS infections. However, beyond the intrinsic quality of a given molecular tool, we should bear in mind that it is just a tool, albeit an incredibly ingenious one. Its interest in clinical practice and for the patient's benefit is related to its proper use, with the appropriate specimen and at the appropriate moment in the course of the disease and with the supervision of pre- and postanalytic steps to achieve a result within the time frame of clinical decision-making. Molecular testing should be discipline-based and not instrument-based. The final assessment is based on a number of different clues that the physician considers in their entity to make a judgment.

12.6 Take-home messages

– In CNS infectious disease, CSF is the reference specimen. A sufficient volume of CSF is crucial for extensive investigations and allows multiple specimens to be stored in suitable conditions. Laboratories should receive full clinical information, especially about therapy before lumbar puncture.
– For bacterial meningitis, vaccination has an impact on the occurrence and transmission of the disease, which may change over time and between countries. *Streptococcus pneumoniae* has emerged as the most common cause of community-acquired bacterial meningitis in immunocompetent patients, followed by *Neisseria meningitidis*. Suspicion of bacterial meningitidis has to be considered as an emergency and the patient hospitalized. Clinical diagnosis may be difficult. Lumbar puncture should be widely performed. Early diagnosis and effective antibiotic treatment are the conditions of successful management and prognosis. Detection of *Neisseria meningitidis* and *Streptococcus pneumoniae* DNAs can be successfully performed in emergencies. Sensitivities of molecular techniques for *Neisseria meningitidis*, *Streptococcus pneumoniae*, and *Haemophilus influenzae* are worse than cultures. However, they may add additional positives when antibiotics are given before lumbar puncture. Culture of CSF remains the gold standard for confirming etiological diagnosis and testing susceptibility to antibiotics. CNS tuberculosis is probably more frequent than traditionally reported, presenting with atypical features such as rapid course and uncommon age distribution. Morbidity and mortality remain high. Because etiological diagnosis is difficult

because of the poor sensitivity of diagnostic tools, every effort has to be made to consider the possibility of *Mycobacterium tuberculosis* as a cause of CNS disease and to initiate specific therapy in suspected cases. Molecular detection of *Mycobacterium tuberculosis* needs to be improved.

– For viral meningitis, it must be considered that approximately 5% of patients may have unremarkable CSF standard investigations. Molecular tests in viral CNS infection have become the gold standard. EV meningitis is by far the most frequent. It occurs not only in summer and early fall but also during winter and is common in adults. The differential clinical diagnosis on admission is difficult. However, in almost all cases the disease is benign with spontaneous rapid recovery. Molecular diagnosis should be performed systematically. EV CNS disease remains an important cause of morbidity and financial burden and merits efforts to improve diagnostics, treatment, and prevention options. Human parechoviruses must be considered as possible cause of CNS infections, especially in the younger. These viruses must be better and more frequently investigated using molecular assays.

– Of the numerous agents that can cause encephalitis, the majority are viruses. Establishing an etiological diagnosis of a CNS viral infection may be difficult. Epidemiology of various pathogens has changed in recent years. The first organisms to be sought for should be those that are susceptible to specific treatment. Empirical antimicrobial agents should be used immediately when specific pathogens, for example HSV-1, HSV-2, or VZV, are suspected. Molecular diagnosis in CSF is now the gold standard for herpesvirus infections. However, if a molecular test reveals undetectable herpesvirus DNA, it must be repeated in cases of persistent suspicion.

12.7 Acknowledgment

The author thanks Audrey Mirand and Christine Archimbaud for excellent technical help.

12.8 Further reading

[1] Archimbaud, C., Ouchane, L., Mirand, A., Chambon, M., Demeocq, F., Labbe, A., Laurichesse, H., Schmidt, J., Clavelou, P., Aumaître, O., Regagnon, C., Bailly, J.L., Henquell, C., Peigue-Lafeuille, H. (2013) Improvement of the management of infants, children and adults with a molecular diagnosis of enterovirus meningitis during two observational study periods. PLoS One 8:e68571.

[2] Chaudhuri, A., Martinez-Martin, P., Kennedy, P.G., Andrew Seaton, R., Portegies, P., Bojar, M., Steiner, I., EFNS Task Force (2008) EFNS guideline on the management of community-acquired bacterial meningitis: report of an EFNS Task Force on acute bacterial meningitis in older children and adults. Eur. J. Neurol. 15:649–659.

[3] Christie, L.J., Loeffler, A.M., Honarmand, S., Flood, J.M., Baxter, R., Jacobson, S., Alexander, R., Glaser, C.A. (2008) Diagnostic challenges of central nervous system tuberculosis. Emerg. Infect. Dis. 14:1473–1475.

[4] Eichner, M., Brockmann, S.O. (2013) Polio emergence in Syria and Israel endangers Europe. Lancet 382:1777.

[5] Glaser, C.A., Honarmand, S., Anderson, L.J., Schnurr, D.P., Forghani, B., Cossen, C.K., Schuster, F.L., Christie, L.J., Tureen, J.H. (2006) Beyond viruses: clinical profiles and etiologies associated with encephalitis. Clin. Infect. Dis. 43:1565–1577.

[6] Harvala, H., Wolthers, K.C., Simmonds, P. (2010) Parechoviruses in children: understanding a new infection. Curr. Opin. Infect. Dis. 23:224–230.

[7] Luo, H.M., Zhang, Y., Wang, X.Q., Yu, W.Z., Wen, N., Yan, D.M., Wang, H.Q., Wushouer, F., Wang, H.B., Xu, A.Q., Zheng, J.S., Li, D.X., Cui, H., Wang, J.P., Zhu, S.L., Feng, Z.J., Cui, F.Q., Ning, J., Hao, L.X., Fan, C.X., Ning, G.J., Yu, H.J., Wang, S.W., Liu, D.W., Wang, D.Y., Fu, J.P., Gou, A.L., Zhang, G.M., Huang, G.H., Chen, Y.S., Mi, S.S., Liu, Y.M., Yin, D.P., Zhu, H., Fan, X.C., Li, X.L., Ji, Y.X., Li, K.L., Tang, H.S., Xu, W.B., Wang, Y., Yang, W.Z. (2013) Identification and control of a poliomyelitis outbreak in Xinjiang, China. N. Engl. J. Med. 369:1981–90.

[8] Mailles, A., Stahl, J.P., Steering Committee and Investigators Group (2009) Infectious encephalitis in France in 2007: a national prospective study. Clin. Infect. Dis. 49:1838–1847.

[9] Mansfield, K.L., Johnson, N., Phipps, L.P., Stephenson, J.R., Fooks, A.R., Solomon, T. (2009) Tick-borne encephalitis virus – a review of an emerging zoonosis. J. Gen. Virol. 90:1781–1794.

[10] Mirand, A., Archimbaud, C., Chambon, M., Regagnon, C., Brebion, A., Bailly, J.L., Peigue-Lafeuille, H., Hemquell, C. (2012) Diagnosis of human parechovirus infections of the central nervous system with a commercial real-time reverse transcription-polymerase chain reaction kit and direct genotyping in cerebrospinal fluid specimens. Diagn. Microbiol. Infect. Dis. 74:78–80.

[11] Mirand, A., Schuffenecker, I., Henquell, C., Billaud, G., Jugie, G., Falcon, D., Mahul, A., Archimbaud, C., Terletskaia-Ladwig, E., Diedrich, S., Huemer, H., P., Enders, M., Lina, B., Peigue-Lafeuille, H., Bailly, J.L. (2010) Phylogenetic evidence for a recent spread of two populations of human enterovirus 71 in Europe countries. J. Gen. Virol. 91:2263–2277.

[12] Rock, R.B., Olin, M., Baker, C.A., Molitor, T.W., Peterson, P.K. (2008) Central nervous system tuberculosis: pathogenesis and clinical aspects. Clin. Microbiol. Rev. 21:243–261.

[13] Tunkel, A.R., Glaser, C.A., Bloch, K.C., Sejvar, J.J., Marra, C.M., Roos, K.L., Hartman, B.J., Kaplan, S.L., Scheld, W.M., Whitley, R.J., Infectious Diseases Society of America (2008) The management of encephalitis: clinical practice guidelines by the Infectious Diseases Society of America. Clin. Infect. Dis. 47:303–327.

[14] Tyler, K.L. (2009) Emerging viral infections of the central nervous system: part 1. Arch. Neurol. 66:939–948.

[15] Venkatesan, A., Tunkel, A.R., Bloch, K.C., Lauring, A.S., Sejvar, J., Bitnun, A., Stahl, J.P., Mailles, A., Drebot, M., Rupprecht, C.E., Yoder, J., Cope, J.R., Wilson, M.R., Whitley, R.J., Sullivan, J., Granerod, J., Jones, C., Eastwood, K., Ward, K.N., Durrheim, D.N., Solbrig, M.V., Guo-Dong, L., Glaser, S.A., on behalf of the International Encephalitis Consortium (2013) Case definitions, diagnostic algorithms, and priorities in encephalitis: consensus statement of the International Encephalitis Consortium. Clin. Infect. Dis. 57:1114–1128.

13 Pathogens relevant in sexually transmitted infections

Suzanne M. Garland and Sepehr N. Tabrizi

Sexually transmitted infections (STIs) are common, particularly amongst those under 25 years of age. They are diverse in clinical presentation, many of which are asymptomatic, yet have the potential to cause significant morbidity and mortality if they go unrecognized and untreated. In particular, STIs in the pregnant woman can result in adverse outcomes for the pregnancy, as well as produce congenital abnormalities or disease in the fetus or newborn.

Pathogens include viruses, as well as bacteria, protozoa, and parasites. Definitive diagnoses in some instances have been difficult because pathogens such as the human papillomavirus (HPV) may not be propagated by conventional culture methods. Conventional diagnoses of others such as *Neisseria gonorrhoeae* and *Haemophilus ducreyi* have been hampered by transport requirements and necessity for specialized media due to the fastidious nature of the organisms.

With the advent of molecular techniques, diagnoses have become more readily available. However, these techniques require careful collection, transport and handling at the clinical level, as well as in the laboratory to ensure that there is no cross-contamination. Molecular techniques, being more sensitive and rapid (results within hours of receipt in the laboratory), offer great advantages over traditional diagnoses, although are generally more expensive, require scientific expertise, and currently do not give antimicrobial sensitivity results. In addition, as they are detecting DNA or RNA, the results need to be interpreted according to the clinical setting, as they may represent, for example, detection of latent virus in the setting of a past herpes infection and not be the cause of the particular lesion in question; or DNA may be detected following a recently and adequately treated STI infection and represent "dead organisms". This may cause confusion in interpretation followed by overtreatment. In screening for some STIs, molecular assays allow for detection of pathogens in self-collected specimens, which has advantages for difficult to access populations, as such samples can be sent to the laboratory for processing.

The human immunodeficiency virus (HIV) and the hepatitis B virus (HBV) have been included in other chapters (see Chapters 7 and 8).

13.1 Symptoms and clinical manifestations

Table 13.1 outlines various viral pathogens associated with STIs, denoting the areas of infection as well as symptoms and common clinical signs. It needs to be borne

Tab. 13.1: Viral pathogens in STIs.

Virus	Area of infection, transmission pattern	Symptoms and clinical manifestations
Adenoviruses	Respiratory tract, gastrointestinal tract, eyes, bladder: specifically urethra, throat. Transmitted through saliva, eye and genital secretions.	Urethritis: urinary frequency, dysuria or discharge (also pharyngitis, gastroenteritis, keratoconjunctivitis).
CMV	Respiratory tract, liver, genital tract. Transmitted through saliva, urine, genital secretions, blood (viremia).	Largely asymptomatic, fever, nonspecific skin rash, cervical lymphadenopathy (occasionally hepatitis).
HSV	Genital skin and mucous membranes and sensory nerve root ganglia: oropharynx, uncommonly viremia. Transmitted through saliva, genital secretions.	Genital or other mucocutaneous vesicles or ulcers, genital discharge, fever general malaise, myalgia, headache (uncommonly meningitis, encephalitis, hepatitis).
HPV	Infects genital skin and mucus membranes. Low-risk genotypes (6, 11) cause the majority of genital warts or condylomata accuminata. In addition, they may cause (uncommonly) recurrent respiratory papillomatosis. High risk genotypes (especially 16 and 18) cause cervical, some vulvar, some vaginal and anal dysplasias, as well as their respective cancers. In addition, they may cause a proportion of oropharyngeal cancers (particularly of the tonsil, back of tongue).	Most often asymptomatic. Genital warts, abnormal cervical cytology on Pap screening, abnormal genital bleeding and/discharge in association with neoplasia.

in mind, however, that many STIs are in fact asymptomatic. Table 13.2 outlines the various bacterial and protozoan pathogens associated with STIs, denoting the areas of infection as well as symptoms and clinical manifestations.

Any individual exposed to or diagnosed with an STI should be tested for other STIs relevant to the local epidemiology. In addition, their partner(s) (and in the case of a pregnant woman, their infant and depending on the organism the placenta at delivery) should also be investigated and treated appropriately and simultaneously to avoid "ping-pong" transmission.

Chlamydia (C.) trachomatis is the most common bacterial cause of STI. Over the past decade, notifications have increased more than can be explained by more frequent testing and tests that are more sensitive and relating to sexual behaviors. Those most at risk are young women (under 25 years), particularly those who have recently changed partners.

Tab. 13.2: Bacterial and protozoan pathogens in STIs.

Pathogen	Area of infection, transmission pattern	Symptoms and clinical manifestations
Chlamydia trachomatis	Columnar epithelium of cervix, urethra, Fallopian tubes, conjunctiva of eye, lungs, anal canal. Genital secretions (transmission).	Most often asymptomatic. Genital tract discharge, cervicitis, urethritis, pelvic inflammatory disease (PID), conjunctivitis urinary frequency, abnormal bleeding, sticky eye, pneumonitis (infants with nasopharyngeal infection). Complications include: ectopic pregnancy, infertility, chronic pelvic pain. In the pregnant woman, adverse outcomes include spontaneous abortion, postpartum (or postabortal) endometritis and salpingitis, premature rupture of the membranes and (rarely) intrauterine fetal infection.
Chlamydia trachomatis	Genital epithelium and mucus membranes, anal canal, lymph node. Genital secretions (transmission).	Genital ulceration, lymphatic buboes, proctitis, genital ulceration.
Haemophilus ducreyi	Genital epithelium and mucus membranes. Genital secretions (transmission).	Genital tract discharge, pain, ulceration.
Klebsiella granulomatis	Genital epithelium and mucus membranes. Genital secretions (transmission).	Genital tract discharge, pain, ulceration.
Neisseria gonorrhoeae	Cervix, urethra, Fallopian tubes, eye, rarely joints. Genital secretions (transmission, rarely blood-borne).	Many women and men are asymptomatic. Cervicitis, urethritis PID, conjunctivitis, rarely: septicemia, septic arthritis.
Mycoplasma genitalium	Cervix, urethra and possibly Fallopian tubes, anal canal.	May be asymptomatic. Genital tract discharge, pain, cervicitis, urethritis, possibly PID and proctitis.
Treponema pallidum	Genital epithelium and mucus membranes. Genital secretions and blood-borne (transmission).	May be asymptomatic, genital ulceration, rash, genital ulcerations untreated complications of CVS, CNS.
Trichomonas vaginalis	Genital mucus membranes of vagina and epithelium of the bladder.	May be asymptomatic or genital discharge, vaginitis urinary frequency.

Infection with *Neisseria (N.) gonorrhoeae*, in particular in extra-genital sites (e.g. oral, rectal) are on the increase, in particular among men who have sex with men populations. Overall, because of the poor sensitivity of conventional culture in the diagnosis of gonorrhoea, molecular methods are being utilized. The challenge for molecular detection of *N. gonorrhoeae* is choosing the appropriate target site, given the high natural competence of this organism for transformation during its entire life cycle. This commonly results in exchange of genetic information with other related *N. meningitidis*, as well as with commensal *Neisseria* species. This can generate a problem particularly for extragenital sites such as the pharynx, that frequently contain high numbers of nongonococcal *Neisseria* species. As none of the commercial assays currently have approval for detection of this infection from extragenital sites and because of the importance of detection of this infection in these sites for appropriate treatment, many laboratories, following in-house validation, have implemented testing of such samples. To overcome the specificity issues of molecular assays, confirmation of positive results using supplementary assays for all gonococcal infections is advised and has become routine in many settings worldwide. This is particularly advised when diagnosing gonorrhoea from extra-genital sites, especially in low disease prevalence settings.

The relatively recently recognized *Mycoplasma genitalium* as a pathogen also warrants investigation as part of routine STI screening. Detection of this pathogen can now be readily implemented by various laboratories with molecular assay capabilities; however, there are no commercially available assays currently on the market.

For certain STIs (as well as other infectious diseases), routine antenatal screening is indicated. In particular, this is important where there is a risk of fetal or neonatal infection and/or adverse pregnancy outcome if maternal infection occurs. This applies where screening and confirmatory tests are available, inexpensive, sensitive, and specific, and in addition, safe and effective treatment can reduce morbidity and mortality in the fetus and/or the mother.

An example is that of *Treponema (T.) pallidum*. Despite syphilis being recognized since antiquity, the causative bacterial agent, *T. pallidum* is still unable to be grown by conventional means. Thus the diagnosis is based on positive serological tests, including one type which detects specific antibodies against *T. pallidum* (*T. pallidum* hemagglutination assay) indicating active or past infection and the other for nonspecific reagin-type antibodies (rapid plasma reagin test), both types of which if reactive together, indicate active infection or recently treated active infection. A reactive rapid plasma reagin test alone may be a "biological false-positive" and common during pregnancy; reactive *T. pallidum* antibody alone suggests past, treated or inactive infection. Following treatment for active infection patients should be followed up by nonspecific tests to ensure that the rapid plasma reagin-test titer falls significantly when sera are tested in parallel.

13.2 Preanalytics

Tables 13.3 and 13.4 summarize suitable specimens to test for each analyte including appropriate handling.

13.3 Analytics

Following specimen collection, nucleic acid from samples must be extracted. A variety of manual and automated methods for extraction is available. Adequate measures should be put in place to ensure extraction has taken place efficiently.

Detection of an adequate quantity of a human housekeeping gene such as beta-globin has the advantage of indicating whether an appropriate sample has been collected. This is particularly important in cases of self-collected samples.

For amplification of nucleic acids, well-verified laboratory developed assays can be used in routine diagnostics; however, where an approved commercial assay is available, it should be utilized. Use of commercial assays is advantageous, as they usually include quality controlled reagents and appropriate controls. Some currently available commercial assays are listed in Tab. 13.5 and Tab. 13.6. As indicated above, confirmation of positive samples is important for some pathogens such as *N. gonorrhoeae*, in particular if the sample is from rectal or pharyngeal sites, and by including supplemental assays targeting another genomic region as a crucial step.

Tab. 13.3: Appropriate specimen collection and handling for viral STIs.

Virus	Biological specimen	Handling
Adenoviruses	Urethral swab or urine	Urethral swab can be transported dry or in viral transport medium. First void urine should be collected in a sterile jar. Urine specimen jars are prone to leakage and particular attention needs to be paid to ensure jars are sealed prior to transportation. Transport at room temperature within 24 h, or up to 4 days at 4°C.
HSV	Swab of ulcer or vesicular lesion (ensure vesicle broken before swabbing the base of the lesion), serum (serology)	Swab can be transported dry or in viral transport medium. Transport at room temperature within 24 h or up to 4 days at 4°C.
HPV	Swab, scrape, tissue biopsy	Commercial collection systems available. Swabs, scrapes, and biopsies should be transported in those or in viral transport medium. Transport conditions according to the manufacturer's package insert. If viral transport medium used, transport at room temperature within 24 h or up to 4 days at 4°C.

Tab. 13.4: Appropriate specimen collection and handling for bacterial and protozoon STIs.

Pathogen	Biological specimen	Handling
Chlamydia trachomatis	Urethral, cervical, high vaginal, rectal and pharyngeal swab, urine (rarely bubo aspirate, Fallopian tube brushings)	Commercial collection systems available. Swabs should be transported in those or in viral transport medium. First void urine should be collected in sterile jar. Transport conditions according to the manufacturer's package insert. If viral transport medium used, transport at room temperature within 24 h or up to 4 days at 4°C.
Haemophilus ducreyi	Urethral, high vaginal, genital ulcer swab	Swab can be transported dry or in viral transport medium. First void urine should be collected in sterile jar. Transport at room temperature within 24 h or up to 4 days at 4°C.
Klebsiella granulomatis	Urethral, high vaginal, genital ulcer swab	Swab can be transported dry or in viral transport medium. Transport at room temperature within 24 h or up to 4 days at 4°C.
Neisseria gonorrhoeae	Urethral, cervical, rectal and pharyngeal swab urine (occasionally joint aspirate or blood culture)	Commercial collection systems available. Swabs should be transported in those or in viral transport medium. First void urine should be collected in sterile jar. Transport conditions according to the manufacturer's package insert. If viral transport medium used, transport at room temperature within 24 h or up to 4 days at 4°C.
Mycoplasma genitalium	Urethral, cervical, high vaginal swab and urine	Swab can be transported dry or in viral transport medium. First catch urine should be collected in sterile jar. Transport at room temperature within 24 h or up to 4 days at 4°C.
Treponema pallidum	Urethral and high vaginal swab, serum or plasma	Swab can be transported dry or in viral transport medium. Separate serum or plasma within 24 h. Transport at room temperature within 24 h or up to 4 days at 4°C.
Trichomonas vaginalis	Urethral and high vaginal swab, urine	Swab can be transported dry or in viral transport medium. First catch urine should be collected in sterile jar. Transport at room temperature within 24 h or up to 4 days at 4°C.

Tab. 13.5: Target nucleic acids, methods, and major commercial viral STI assays.

Virus	Target nucleic acid	Recommended method and currently available commercial assays
Adenoviruses	DNA	Mainly laboratory-developed assays based on qPCR.
CMV	DNA	See Chapter 9 (Tab. 9.4).
HSV	DNA	See Chapter 12 (Tab. 12.4).
HPV	DNA, mRNA	Abbott RealTime High Risk HPV (qPCR), Roche COBAS 4800 HPV Test (qPCR), Qiagen Digene HPV HC2 DNA Test (hybrid capture), Genomica CLART Human Papillomavirus (PCR and microarray), Greiner Bio-One PapilloCheck (PCR and microarray), Innogenetics INNO-LiPA HPV Genotyping Extra (PCR and line probe assay), Roche Linear Array HPV (PCR and line probe assay), NucliSENS EasyQ HPV (NASBA). Hologic APTIMA HPV (TMA), Hologic Cervista HPV HR (Invader assay)

Tab. 13.6: Target nucleic acids, methods and major commercial bacterial and protozoon STI assays.

Pathogen	Target nucleic acid	Recommended method and currently available commercial assays
Chlamydia trachomatis	DNA	Abbott RealTime CT/NG (qPCR), Roche COBAS 4800 System CT/NG Test (qPCR), Roche COBAS TaqMan CT Test v2.0 (qPCR), Siemens VERSANT CT/GC DNA 1.0 Assay (qPCR), BD ProbeTec ET CT/GC Assay (SDA), Gen-Probe APTIMA Combo 2 Assay (TMA), Cepheid GenXpert CT/NG (qPCR).
Haemophilus ducreyi	DNA	Laboratory-developed qPCR.
Klebsiella granulomatis	DNA	Laboratory-developed qPCR.
Neisseria gonorrhoeae	DNA	Abbott RealTime CT/NG (qPCR), Roche COBAS 4800 System CT/NG Test (qPCR), Siemens VERSANT CT/GC DNA 1.0 Assay (qPCR), BD ProbeTec ET CT/GC Assay (SDA), Gen-Probe APTIMA Combo 2 Assay (TMA). Cepheid GenXpert CT/NG (qPCR). NOTE: Due to cross reaction with commensal strains, it is imperative that positive results be tested by an alternative supplemental assay.
Mycoplasma genitalium	DNA	Laboratory-developed qPCR.
Treponema pallidum	DNA	Laboratory-developed qPCR.
Trichomonas vaginalis	DNA	Aptima Gen-Probe (TMA).

Monitoring of the positivity rate in the laboratory can also highlight problems with the ability of a particular assay to detect the intended target. This was highlighted with a variant of *C. trachomatis* circulating in Sweden, which had a deletion spanning a target of some commercial assays. This caused the assays to result in false-negatives and in non-treatment of infected individuals. Therefore, monitoring of decreases or for increases in prevalence is an important measure to detect such issues with assays in use or false-positives, respectively.

13.4 Postanalytics

Results must be analyzed and appropriately interpreted (Tab. 13.7). For qPCR based tests, control results must fall within the predetermined crossing threshold. Results should be interpreted as positive or negative or a quantitative result is obtained by comparison with a standard curve of known copy number. Samples that have inadequate internal control amplification and negative target result should be either repeated or reported as not assessable. Participation in proficiency panels and quality assurance programs is essential to determine assay performance and every effort must be made to participate in such programs for each analyte.

Detection of pathogens by molecular methods generally indicates the presence of an organism's nucleic acid; hence, this could be nonviable, post-treatment, or

Tab. 13.7: Reporting of results.

Pathogen	Result	Recommendation
Adenoviruses	Detected/not detected	
HSV	Detected/not detected	HSV type 1 and type 2 may be discriminated.
HPV	Detected/not detected	Genotyping with high-risk types may have clinical impact on follow-up and treatment of patients.
Chlamydia trachomatis	Detected/not detected (genotype may be warranted for surveillance)	
Chlamydia trachomatis	Detected/not detected	
Haemophilus ducreyi	Detected/not detected	
Klebsiella granulomatis	Detected/not detected	
Neisseria gonorrhoeae	Detected/not detected	Confirmation of positives required.
Mycoplasma genitalium	Detected/not detected	
Treponema pallidum	Detected/not detected	
Trichomonas vaginalis	Detected/not detected	

represent latent infection such as in the case of herpes viruses and not necessarily the cause of a particular episode of clinical disease. Therefore, the results should generally be interpreted along with the clinical presentation.

13.5 Further reading

[1] Bowden, F.J., Tabrizi, S.N., Garland, S.M., Fairley, C.K. (2002) Infectious diseases. 6: Sexually transmitted infections: new diagnostic approaches and treatments. Med. J. Aust. 176:551–557.

[2] Chernesky, M.A. (2005) The laboratory diagnosis of *Chlamydia trachomatis* infections. Can. J. Infect. Dis. Med. Microbiol. 16:39–44.

[3] Coutlée, F., Rouleau, D., Ferenczy, A., Franco, E. (2005) The laboratory diagnosis of genital human papillomavirus infections. Can. J. Infect. Dis. Med. Microbiol. 16:83–91.

[4] Garland, S.M., Tabrizi, S.N. (2004) Diagnosis of sexually transmitted infections (STI) using self-collected non-invasive specimens. Sex. Health 1:121–126.

[5] Garland, S.M., Tabrizi, S.N. (2006) HPV DNA testing: potential clinical applications. Int. J. Health Prom. Educ. 44:107–112.

[6] Gilbert, G., O'Reilly, M., Garland, S.M. (2010) Infectious diseases in pregnancy and the newborn. In: Yung, A., McDonald, M., Spelman, D., Street, A., Johnson, P., Sorrell, T., McCormack, J., eds. Infectious diseases: a clinical approach, 3rd Ed., East Hawthorn, Victoria: IP Communications.

[7] Tabrizi, S.N., Unemo, M., Limnios, A.E., Hogan, T.R., Hjelmevoll, S.O., Garland, S.M., Tapsall, J. (2011) Evaluation of six commercial nucleic acid amplification tests for the detection of *Neisseria gonorrhoeae* and other Neisseria species. J. Clin. Microbiol. 49:3610–3615.

[8] Twin, J., Taylor, N., Garland, S.M., Hocking, J.S., Walker, J., Bradshaw, C.S., Fairley, C.K., Tabrizi, S.N. (2011) Comparison of two *Mycoplasma genitalium* real-time PCR detection methodologies. J. Clin. Microbiol. 49:1140–1142.

[9] Unger, E.R., Dillner, J. (eds.) (2009) Human papillomavirus laboratory manual. First ed., WHO Department of Immunization, Vaccines and Biologicals.

[10] Whiley, D.M., Garland, S.M., Harnett, G., Lum, G., Smith, D.W., Tabrizi, S.N., Sloots, T.P., Tapsall, J.W. (2008) Exploring 'best practice' for nucleic acid detection of Neisseria gonorrhoeae. Sex. Health 5:17–23.

Index

Acceptor molecule, 76
Acceptor probe, 68
Accreditation, 47
Accuracy, 50
Acquired immunodeficiency syndrome. *See*
    AIDS
Additive, 30
Adenovirus, 1, 125, 129, 132, 140, 147, 157
– Clinical presentation, 1
– Complications, 1
– Epidemiology, 1
– Incubation period, 1
– qPCR, 132
– Reactivation, 140
– Sample material, 1
– Serotypes, 132
– Transmission, 1
Agreement, 50
AIDS, 97, 99, 100
Amplification, 64, 70
– Efficiency, 45
– Isothermal, 75
Amplification efficiency, 72
Amplification product, 67, 69
– Sequencing, 86
Analyte, 42, 49, 50, 53, 54
Analytical measuring range, 72, 89, 96
Anaplasma phagocytophilum, 196
– Nested PCR, 207
Annealing, 68, 69
Archival material, 80
Aspergillus, 25
– Clinical presentation, 26
– Complications, 26
– Epidemiology, 25
– Sample material, 26
– Transmission, 25
Asthma, 147
Astrovirus, 2
– Clinical presentation, 2
– Epidemiology, 2
– Incubation period, 2
– Sample material, 2
– Transmission, 2
Ataxia, 195
Automation, 58, 67

bDNA, 80
– Sensitivity, 80
Biopsy, 85
BKPyV, 14, 85, 127, 131, 136, 142
– Clinical presentation, 15
– Epidemiology, 14
– Incubation period, 15
– qPCR, 136
– Reactivation, 142
– Sample material, 15
– Transmission, 14
Blood
– Sampling, 29
Bocavirus, 2, 148, 151, 172
– Clinical presentation, 2
– Complications, 2
– Epidemiology, 2
– Incubation period, 2
– Sample material, 2
– Transmission, 2
Bordetella, 19
– Clinical presentation, 19
– Complications, 19
– Epidemiology, 19
– Sample material, 19
– Transmission, 19
Bordetella bronchiseptica, 146, 156
Bordetella holmesii, 157
Bordetella parapertussis, 146, 156
Bordetella pertussis, 146, 150, 151, 156
Borrelia burgdorferi, 19, 196
– Clinical presentation, 20
– Epidemiology, 19
– Incubation period, 19
– Nested PCR, 207
– Sample material, 20
– Transmission, 19
Bronchiolitis, 147
Bronchoalveolar lavage, 149

Calibration, 44
Campylobacter, 179
Candida, 26
– Clinical presentation, 26
– Complications, 26
– Epidemiology, 26

– Sample material, 26
– Transmission, 26
Capture, 58
Category A Infectious Substances, 34
Category B Infectious Substances, 35
cDNA, 69, 73, 75
CE-labeled, 47
Central nervous system. *See* CNS
Cerebrospinal fluid. *See* CSF
Chemiluminescence, 81
Chlamydia pneumoniae, 151
Chlamydia trachomatis, 20, 38, 76, 218, 224
– Clinical presentation, 20
– Collection, 38
– Complications, 20
– Epidemiology, 20
– Incubation period, 20
– Sample material, 20
– Storage, 39
– Swab, 38
– Transmission, 20
Chlamydophila pneumoniae, 20, 146, 150, 155, 158, 169, 170
– Clinical presentation, 20
– Epidemiology, 20
– Incubation period, 20
– Sample material, 21
– Transmission, 20
Citrate tube, 30
Clostridium difficile, 21
– Clinical presentation, 21
– Complications, 21
– Epidemiology, 21
– Risk factors, 21
– Sample material, 21
– Transmission, 21
CMV, 3, 85, 125, 129, 132, 140
– Clinical presentation, 3
– Complications, 3
– EDTA whole blood, 129
– Epidemiology, 3
– Incubation period, 3
– Latent virus, 129
– Monitoring, 140
– Mutations, 140
– Neutropenia, 133
– Plasma, 129
– qPCR, 132
– Reactivation, 133

– Sample material, 3
– Transmission, 3
– Variant, 133
– Viral load, 140
CNS, 201
– Specimen, 201
CNS infection, 201
– Treatment, 201
Coefficient of variation, 51
Colitis, 178
Colon, 178
Commutability, 43
Competition, 45
Competitive PCR, 88
Consistency, 48
Contamination, 33, 57, 58, 185
Coronavirus, 2, 148, 151, 172
– Clinical presentation, 3
– Epidemiology, 3
– Incubation period, 3
– Sample material, 3
– Transmission, 3
Correctness, 46
Coxsackievirus, 193
Cross contamination, 30
Croup, 147
Cryptosporidium, 185, 190
CSF, 202
– EBV, 210
– HHV6, 211
– Multiplex PCR, 207, 208
– Real-time PCR, 204
– Storage, 204
– Transport, 203
Cytomegalovirus. *See* CMV

Degradation, 59, 210
– DNA, 30
– RNA, 31
Dehydration, 178
Dengue virus, 3
– Clinical presentation, 4
– Complications, 4
– Epidemiology, 3
– Incubation period, 4
– Sample material, 4
– Transmission, 3
Detection, 64
Detector probe, 76

Diarrhea, 175, 176, 178
Digital PCR. *See* dPCR
Dimer, 68, 69, 70
DNA polymerase, 32
DNase, 29, 30
– Inactivation, 31
– Inhibition, 29
Documentation, 48
Donor probe, 68
dPCR, 74
– Accuracy, 74
– Efficiency, 75
– Precision, 74
– Sensitivity, 74

EBV, 4, 85, 125, 130, 133, 141
– Clinical presentation, 5
– Complications, 5
– Epidemiology, 4
– Incubation period, 5
– qPCR, 133
– Sample material, 5
– Sequence variation, 133
– Transmission, 4
Echovirus, 193
EDTA plasma, 37, 38
EDTA whole blood, 30, 37
Ehrlichia chaffeensis, 196
– qPCR, 207
Elongation, 69
Elution, 58
Employee competency, 44, 48
Encephalitis, 193, 195, 196, 200, 201
Encephalomyelitis, 195
Enterococcus, 21
– Clinical presentation, 21
– Complications, 21
– Epidemiology, 21
– Sample material, 21
Enterovirus, 4, 76, 127, 157, 197, 201,
    211, 214
– Clinical presentation, 4
– Complications, 4
– Epidemiology, 4
– Incubation period, 4
– RT-qPCR, 204
– Sample material, 4
– Serotype, 193
– Transmission, 4

Enterovirus 71, 193, 212
Epstein-Barr virus. *See* EBV
Exempt patient specimen, 36
Extension, 70
External run control, 46

FDA-approved, 46, 48, 53
FDA-cleared, 46, 48, 53
feces, 178, 179, 181, 183, 185, 190
– extraction, 181
– inhibitors, 183
– real time PCRs, 186
– storage, 179
– transport, 179
Fluorescein, 68, 69
Fluorescence, 68, 69, 70
Fluorescent, 69
Fluorescent resonance energy transfer.
    See FRET
Freeze-thaw cycle, 37, 38
FRET, 69

Gastroenteritis, 175, 176, 178, 181, 183, 190
– Symptoms, 178
Gel separator, 37, 38
Generic extraction protocol, 61
Genome, 64
Genotype, 43, 70
Giardia, 179, 185, 190
Group B *Streptococcus*, 24
– Clinical presentation, 24
– Complications, 25
– Epidemiology, 24
– Incubation period, 24
– Sample material, 25
– Transmission, 24

HAART, 100
Haemophilus ducreyi, 217
Haemophilus influenzae, 196, 212, 213
– qPCR, 207
Hairpin, 70, 76
Hantavirus, 5
– Clinical presentation, 5
– Epidemiology, 5
– Incubation period, 5
– Sample material, 5
– Transmission, 5
Harmonization, 54

HAV, 5, 113, 117, 121
- Clinical presentation, 6
- Epidemiology, 5
- Incubation period, 5
- Sample material, 6
- Stool, 117
- Transmission, 5
HBV, 6, 37, 43, 85, 113, 115, 117, 121
- Clinical presentation, 6
- Complications, 6
- Drug resistance, 118
- Epidemiology, 6
- Genotype, 116, 117
- Incubation period, 6
- Monitoring, 117
- Sample material, 6
- Sequencing, 118
- Storage, 37
- Transmission, 6
- Viral load, 117
HCV, 6, 38, 43, 76, 85, 113, 118, 122
- Clinical presentation, 6
- Complications, 6
- Confirmation, 122
- Drug resistance, 119
- Epidemiology, 6
- Genotype, 116, 119
- Incubation period, 6
- Monitoring, 118
- Sample material, 7
- Sequencing, 119
- Storage, 38
- Therapy, 118
- Transmission, 6
- Treatment duration, 123
HDV, 7, 113, 119, 122
- Clinical presentation, 7
- Epidemiology, 7
- Incubation period, 7
- Sample material, 7
- Transmission, 7
Helicobacter pylori, 22
- Clinical presentation, 22
- Complications, 22
- Epidemiology, 22
- Sample material, 22
- Transmission, 22
Hemagglutinin, 147
Heparin, 114

Hepatitis, 113, 115, 116
- Adenovirus, 113, 116
- EBV, 113
- Hemorrhagic fever viruses, 121
- Herpesviruses, 113, 120
- Viruses, 113
Hepatitis A virus. *See* HAV
Hepatitis B virus. *See* HBV
Hepatitis C virus. *See* HCV
Hepatitis D virus. *See* HDV
Hepatitis E virus. *See* HEV
Hepatocellular carcinoma, 117
Herpes simplex virus. *See* HSV
Herpesvirus, 128, 201, 225
- Active infection, 128
- Latent infection, 128
HEV, 7, 113, 120, 121
- Clinical presentation, 7
- Epidemiology, 7
- Incubation period, 7
- Sample material, 7
- Transmission, 7
HHV-6, 8, 126, 130, 136, 141
- Clinical presentation, 8
- Complications, 9
- Epidemiology, 8
- Incubation period, 8
- Plasma, 130
- qPCR, 136
- Sample material, 9
- Serum, 130
- Transmission, 8
- Viral load, 141
HHV-7, 9, 127
- Clinical presentation, 9
- Complications, 9
- Epidemiology, 9
- Incubation period, 9
- Sample material, 9
- Transmission, 9
HHV-8, 9, 126, 130, 136, 141
- Clinical presentation, 10
- EDTA whole blood, 130
- Epidemiology, 9
- Incubation period, 10
- Monitoring, 141
- Plasma, 130
- Sample material, 10
- Transmission, 9

Highly active antiretroviral therapy.
  See HAART
High-resolution melting, 70
HIV, 10, 36, 43, 76, 85, 97, 98, 99, 100, 101, 102,
    103, 104, 105, 106, 107, 108, 109, 110, 111
– Acute infection, 99
– CCR5, 97, 103, 109
– Clinical presentation, 10
– Coreceptor, 103, 108
– CXCR4, 97, 98, 103, 108, 109
– DNA, 101, 103, 105
– Donor, 102, 104
– Dried spot, 101
– Drug resistance, 103, 106
– EDTA plasma, 101
– Epidemiology, 10
– Fitness, 111
– Genotyping, 106, 108
– Group, 97, 105, 110
– HAART, 110
– Hybridization, 106
– Incubation period, 10
– Insemination, 102
– Integrase, 97, 103, 105
– Latency, 99
– Monitoring, 102, 110
– NASBA, 104, 105
– Neonates, 101, 110
– NGS, 109
– PBMC, 103, 105
– PCR, 103, 105
– Phenotyping, 107, 109
– Plasma, 101, 102, 103, 111
– Protease, 97, 103
– qPCR, 104
– Quantification, 105, 110
– Recombinant virus, 97, 109
– Reservoir, 103, 111
– Reverse transcriptase, 97, 103
– RNA, 101, 102, 103
– RT-PCR, 105
– RT-qPCR, 105
– Sample material, 10
– Screening, 102
– Sequencing, 106, 109, 111
– Specimen, 100
– Storage, 37, 101
– Subtype, 97, 104, 105, 108
– TMA, 104

– Transmission, 10
– Tropism, 103, 108, 111
– Viral load, 102, 110
– Virtual phenotype, 107
Housekeeping genes, 45, 88
HPV, 13, 76, 217
– Clinical presentation, 13
– Epidemiology, 13
– Incubation period, 13
– Sample material, 13
– Transmission, 13
HSV, 8, 76, 127, 201, 211, 214
– Clinical presentation, 8
– Complications, 8
– Epidemiology, 8
– qPCR, 204
– Sample material, 8
– Transmission, 8
HTLV, 11
– Clinical presentation, 11
– Epidemiology, 11
– Incubation period, 11
– Sample material, 11
– Transmission, 11
Human herpes virus 6. *See* HHV-6
Human herpes virus 7. *See* HHV-7
Human herpes virus 8. *See* HHV-8
Human immunodeficiency virus. *See* HIV
Human papillomavirus. *See* HPV
Human T-lymphotropic virus. *See* HTLV
Hybrid capture, 80, 81
– Contamination, 81
– HPV, 81
– Sensitivity, 81
Hybridization, 70, 71, 79, 80
Hybridization probe, 68, 69
Hydrolysis, 69, 70

IC, 44, 53
– Heterologous, 44, 45, 88
– Homologous, 44, 45, 88
Identification number, 34
Immunocapture, 80
Immunosuppression, 85
Imprecision, 50, 51
– Between-run, 50, 51
– Within-run, 50, 51
In Vitro Diagnostic. *See* IVD
Infection

- Acute respiratory, 146, 172
- Community-acquired, 145
- Respiratory, 145
- Respiratory tract, 172
Infection control, 29, 39
Influenza, 147
Influenza A virus, 168
Influenza B virus, 168
Influenza virus, 11, 147, 157
- Clinical presentation, 11
- Complications, 11
- Epidemiology, 11
- Incubation period, 11
- Sample material, 11
- Transmission, 11
Inhibition, 33, 57, 72, 186
Inhibitor, 32, 33, 44
Instrument, 48
- Calibration, 48
- Function checks, 48
- Maintenance, 48
Interferences, 52
Internal control. See IC
International Standard, 43
International Unit, 42
Interpretation, 93, 94
- Analytic, 93
- Clinical, 93
- Raw data, 93
Intestine, 175, 178
Invader, 80
ISO 15189, 41, 54
ISO 9001, 41
- 2008, 44, 54
IVD, 41, 42, 49, 53
- Performance, 49
IVD/CE-labeled, 46
IVD/CE-labeling, 72

JCPyV, 14, 127, 131, 139, 142
- Clinical presentation, 15
- Epidemiology, 14
- Incubation period, 15
- Sample material, 15
- Transmission, 14

Lab-on-a chip, 78, 82
LAMP, 76
- Specificity, 76
Latent infection, 85

Latent virus, 96
Lavage, 85
Legionella pneumophila, 22, 146, 150, 151, 156, 158, 170
- Clinical presentation, 22
- Complications, 22
- Epidemiology, 22
- Incubation period, 22
- Sample material, 22
- Transmission, 22
Legionnaires' disease, 146
Limit of detection. See LOD
Linearity, 50, 51
Listeria monocytogenes, 196
- qPCR, 207
Liver biopsy, 113
Ljungan virus, 195
LLQ, 53, 54, 89
LNA, 68
Locked nucleic acid (LNA), 68
LOD, 53, 54, 89, 96
Loop, 69
Loop-mediated isothermal amplification. See LAMP
Lower limit of quantitation. See LLQ
Lumbar puncture, 202
Lysis, 58

Magnesium, 32
Magnetic glass particle, 58
Measles virus, 11
- Clinical presentation, 12
- Complications, 12
- Epidemiology, 11
- Incubation period, 12
- Sample material, 12
- Transmission, 12
Melting curve, 70
- Peak value, 70
Melting temperature, 68, 70
Meningitis, 193, 195, 196, 197
Meningoencephalitis, 193, 195, 200
Metagenomics, 79
Metapneumovirus, 12, 148, 151
- Clinical presentation, 12
- Epidemiology, 12
- Incubation period, 12
- Sample material, 12
- Transmission, 12
Methicillin-resistant Staphylococcus aureus. See MRSA

Microarray, 71
Mimivirus, 151
MIQE, 73
Molecular beacon, 69
Monitoring, 95, 96
mRNA, 67
MRSA, 24, 78
– Clinical presentation, 24
– Complications, 24
– Epidemiology, 24
– Incubation period, 24
– Sample material, 24
– Transmission, 24
Multipipette, 30
Multiplex PCR, 67, 72
Mumps virus, 12
– Clinical presentation, 12
– Complications, 12
– Epidemiology, 12
– Incubation period, 12
– Sample material, 13
– Transmission, 12
Mycobacterium tuberculosis, 196, 200, 212, 214
– PCR, 207
Mycobacterium tuberculosis complex, 22
– Clinical presentation, 22
– Complications, 22
– Epidemiology, 22
– Incubation period, 22
– Sample material, 23
– Transmission, 22
Mycoplasma genitalium, 220
Mycoplasma pneumoniae, 23, 146, 150, 151, 155,
    158, 169, 170
– Clinical presentation, 23
– Complications, 23
– Epidemiology, 23
– Incubation period, 23
– Sample material, 23
– Transmission, 23

NASBA, 75, 76
Nasopharyngeal aspirate, 150
Nasopharyngeal swab, 150
Native blood, 30
Neisseria gonorrhoeae, 23, 38, 76, 217,
    220, 221
– Collection, 38
– Storage, 38
– Swab, 38

Neisseria meningitidis, 23, 197, 200,
    212, 213
– qPCR, 207
Neisseria species, 23
– Clinical presentation, 23
– Complications, 23
– Epidemiology, 23
– Incubation period, 23
– Sample material, 24
– Transmission, 23
Neuraminidase, 147
Next generation sequencing.
    See NGS
NGS, 78, 81, 109
Nipah virus, 195
– qPCR, 204
Norovirus, 13, 186, 191
– Clinical presentation, 13
– Epidemiology, 13
– Sample material, 13
– Transmission, 13
Nucleic acid
– Release, 58
– Stabilization, 58
Nucleic acid extraction
– Automated, 58
– Eluate, 60
– Elution volume, 59
– Input volume, 59
– Manual, 57
– Platform, 59, 60
– Specimen, 59
Nucleic acid sequence-based amplification.
    See NASBA

Oropharyngeal swab, 150

Packaging, 35
– Label, 35
Papillomavirus. *See* HPV
Parainfluenza virus, 14, 147, 157,
    169, 170
– Clinical presentation, 14
– Epidemiology, 14
– Incubation period, 14
– Sample material, 14
– Transmission, 14
Paralysis, 193, 195
Parechovirus, 195, 214
– qPCR, 205

Parvovirus, 14, 127
– Clinical presentation, 14
– Complications, 14
– Epidemiology, 14
– Incubation period, 14
– Sample material, 14
– Transmission, 14
Patient identification, 29
PCR, 65
– Amplification efficiency, 87
– Array chip, 81
– Bordetella pertussis, 156
– Broad-range, 67
– *Chlamydophila pneumoniae*, 155
– *Legionella pneumophila*, 156
– Multiplex, 67, 72, 158
– *Mycoplasma pneumoniae*, 155
– Plateau effect, 87
– Reliability, 73
– *Streptococcus pneumoniae*, 154
– Urine, 82
Performance characteristics, 90
Pertussis, 146
Ping-pong transmission, 218
Plasmids, 45
PML, 127
Pneumocystis jirovecii, 27
– Clinical presentation, 27
– Complications, 27
– Epidemiology, 27
– Sample material, 27
– Transmission, 27
Pneumonia, 145, 147
– Community-acquired, 145, 146
– Hospital-acquired, 146
Poliovirus, 193
Polymerase chain reaction. *See* PCR
Polyomavirus, 151
Polyomavirus BK. *See* BKPyV
Polyomavirus JC. *See* JCPyV
Polyomavirus-associated hemorrhagic cystitis.
    See PyHC
Polyomavirus-associated nephropathy. *See*
    PyVAN
Post-transplant lymphoproliferative disease..
    See PTLD
Preanalytics, 29
Pre-emptive therapy, 72
Probe, 68

Probit analysis, 54
Proficiency testing, 46, 47
Program, 47
Progressive multifocal leukoencephalopathy.
    See PML
PTLD, 125, 141
Purification, 58
PyHC, 127
PyVAN, 127, 142

Q-LAMP, 76, 77
– Simplicity, 77
qPCR, 66, 67, 68, 72, 73, 82, 85, 88, 113, 115, 224
– Assay setup, 60
– Crossing point, 89
– Cycle threshold, 89
– Fluorescence, 88
– Normalization, 89
– Quantification, 66
– Result, 89
Quality assurance, 41, 46
Quality control, 41, 44, 46
Quality improvement, 41
Quantification, 67, 85
Quantitative loop-mediated isothermal
    amplification. *See* Q-LAMP
Quencher, 69

Rabies virus, 15, 195
– Clinical presentation, 15
– Epidemiology, 15
– Incubation period, 15
– qPCR, 205
– Sample material, 15
– Transmission, 15
Ramification amplification, 80
Reactivation, 72
Real-time PCR. *See* qPCR
Recombinase, 77
Recovery, 50, 52, 54
Reference materials, 41, 42, 44, 46, 47,
    50, 53
Report, 89, 90, 91, 93, 94
– Accession number, 93
– Address, 92
– Bacterial names, 91
– Clinical history, 93
– Date, 93
– Date of birth, 92

– Discrepancy, 93
– Gene name, 91
– General comments, 94
– Genus, 91
– Interpretation, 92
– Nomenclature, 91, 92
– Procedure, 93
– Sensitivity, 90
– Signature, 93
– Significance, 94
– Species, 91
– Specificity, 90
– Specimen source, 92
– Summary, 94
– Target, 93
– Usefulness, 94
Reporter, 69
reproducibility, 53, 89
Respiratory infections
– Viral, 151
Respiratory syncytial virus. *See* RSV
Respiratory tract, 145
– Lower, 145
– Upper, 145
Respiratory tract infections, 39
Result, 89, 91, 92, 93, 94
– Correlation, 92
– False-negative, 44, 49
– False-positive, 49, 53
– Interpretation, 89, 92
– qPCR, 89, 96
– Reference range, 92
– Residual sample, 93
– Significance, 92
– Specimen, 90
– Test name, 92
Retroviridae, 97
Reverse transcription-qPCR. *See* RT-qPCR
Rhinovirus, 16, 147, 157
– Clinical presentation, 16
– Complications, 16
– Epidemiology, 16
– Incubation period, 16
– Sample material, 16
– Transmission, 16
Ribosomal RNA, 86
– 16S rRNA, 86
– 23S rRNA, 86
RNA probe, 80

RNase, 31, 58, 75
– Inactivation, 32
Rotavirus, 16, 191
– Clinical presentation, 17
– Complications, 17
– Epidemiology, 16
– Incubation period, 16
– Sample material, 17
– Transmission, 16
RPA, 77
– Robustness, 78
– Speed, 78
RSV, 15, 147, 157, 168, 170
– Clinical presentation, 16
– Complications, 16
– Epidemiology, 15
– Incubation period, 15
– Sample material, 16
– Transmission, 15
RT-qPCR, 66, 73, 88, 118
Rubella virus, 17
– Clinical presentation, 17
– Complications, 17
– Epidemiology, 17
– Incubation period, 17
– Sample material, 17
– Transmission, 17

Salmonella, 191
Sample
– Low positive, 51
– Negative, 51
– Positive, 51, 52
Sample collection, 29
Sample identity, 34
Sample integrity, 29
Sample matrix, 52
SARS, 172
Scorpion, 70
SDA, 76
Secondary reference materials, 43
Secretion, 85
Seminal fluid, 85
Sensitivity, 72
– Analytical, 186
– Clinical, 49, 186
– Diagnostic, 44, 49
Sepsis, 195, 200
Sequence, 64, 70

Sequencing, 53
– Abundance, 79
– Massive parallel, 78, 81, 109
Serum, 37, 38
Sexually transmitted infection, 217, 218
– Extraction, 221
– Specimen, 221
Shedding, 179, 190
Signal amplification, 65, 79
– Contamination, 79
Signal detection, 80
Single use device, 29
Solution hybridization, 80
Specificity, 53
– Diagnostic, 44, 49, 186
Specimen collection, 33
Sputum, 149
Standardization, 42, 95
Stem hybrid, 69
Stem-loop DNA, 76
Stomach, 175
Stool, 39, 85, 113
– Collection, 39
– Sampling, 30
– Storage, 39
Stopper, 70
Storage, 30, 31, 32, 53
– DNA, 31
– RNA, 32
Strand displacement amplification.
    See SDA
Streptococcus pneumoniae, 25, 145, 148, 154,
    169, 171, 197, 212, 213
– Clinical presentation, 25
– Complications, 25
– Epidemiology, 25
– Incubation period, 25
– qPCR, 208
– Sample material, 25
– Transmission, 25
Swab, 39, 85
SYBR Green, 68

TaqMan probe, 70
Target amplification, 65, 80
– Contamination, 65
Target DNA, 69
Target region, 67
Target RNA, 73, 75

Target sequence, 64, 79
TBEV, 17, 195, 211, 212
– Clinical presentation, 18
– Complications, 18
– Epidemiology, 17
– Incubation period, 17
– qPCR, 205
– Sample material, 18
– Transmission, 17
Test system, 42
Tick-borne encephalitis, 195
Tick-borne encephalitis virus. *See* TBEV
TMA, 75, 76, 115
Toscana virus. *See* TOSV
TOSV, 196, 211, 212
– qPCR, 205
Toxoplasma gondii, 27
– Clinical presentation, 27
– Complications, 27
– Epidemiology, 27
– Incubation period, 27
– Sample material, 27
– Transmission, 27
Traceability, 47
Transcription mediated amplification.
    See TMA
Transfusion medicine, 76
Transplant recipient, 125
– Bone marrow, 125
– Solid organ, 125
Transplantation, 125, 172
Transport, 30, 34
Treatment response, 64
Treponema pallidum, 220

ULQ, 89
Uncertainty, 47
Unit of measurement, 89
Universal PCR, 208
Upper limit of quantitation. *See* ULQ
Urine
– Sampling, 30

Validation, 41, 44, 49, 54
– Component, 44
Variant, 70
Varicella zoster virus. *See* VZV
Verification, 41, 42, 46, 49, 51
– Component, 44, 49, 50, 53

Viability, 67
Viral load, 64, 87, 95
Viral replication, 95
Viral transport medium, 39
– Stability, 39
– Storage, 39
Virus
– Respiratory, 146
VRE. *See Enterococcus*
VZV, 18, 126, 131, 136, 141, 196, 214
– Clinical presentation, 18
– Complications, 18
– EDTA whole blood, 131
– Epidemiology, 18
– Incubation period, 18

– Monitoring, 142
– qPCR, 136
– Sample material, 18
– Transmission, 18
– Viral load, 142

West Nile virus. *See* WNV
WNV, 19, 196, 212
– Clinical presentation, 19
– Complications, 19
– Epidemiology, 19
– Incubation period, 19
– qPCR, 205
– Sample material, 19
– Transmission, 19